D1454788

301 GREAT IDEAS FOR USING TECHNOLOGY

from America's

Most Innovative Small Companies

INTRODUCTION BY NICHOLAS NEGROPONTE

EDITED BY PHAEDRA HISE

Editorial Director: Bradford W. Ketchum, Jr.
Book design: Cynthia M. Davis/Cambridge Prepress

Portions of this book were originally published in *Inc.* and *Inc. Technology* magazines. For information about purchasing back issues of *Inc.* and *Inc. Technology,* please call 617-248-8426.

This book may be purchased in bulk at discounted rates for sales promotions, premiums, or fund raising. Custom books and book excerpts of this publication are available. Contact:
Inc. Business Resources
Attn: Custom Publishing Sales Dept.
38 Commercial Wharf
Boston, MA 02110-3883
1-800-394-1746.

Library of Congress Catalog Number: 97-78371

ISBN 1-880394-77-4

First Edition

www.inc.com/products

CONTENTS

301 GREAT IDEAS FOR USING TECHNOLOGY

Technology befuddles and derails small-business owners more than it has any right to. It's so easy to sink a ton of cash and valuable time into researching, setting up, and managing telephones, computers, networks, and other high-tech systems. For that reason, many small-business owners turn their backs on technology, leaving both its costs and its benefits to their larger competitors.

But they are making a mistake that we hope to help rectify. Many small-business owners have mastered technology and leveraged it to push their companies into new and profitable markets. In this book, we have written about those companies and what their owners have learned. The advice ranges from the mundane—how to save money on purchases—to the arcane—how to structure sales databases.

Compiling these tips was a complex project made easier by the invaluable assistance of many helpful and skilled people. Most notably, thanks are due to the many business owners and managers who took time to talk to *Inc.* writers and editors. We are forever indebted to you for stepping away from the demands of your companies to accept our telephone calls and visits.

I am grateful to the *Inc., Inc. Technology,* and Inc. Online writers and editors who tirelessly gathered and produced the stories. They include Margherita Altobelli, Marc Ballon, Matthew Berk, Alessandra Bianchi, Leslie Brokaw, Leigh Buchanan, Christopher Caggiano, Susan Donovan, Donna Fenn, Jill Andresky Fraser, Robina Gangemi, George Gendron, Susan Greco, Stephanie Gruner, Michael Hofman, Michael Hopkins, Joshua Hyatt, Michelle Keyo, Nancy Lyons, Joshua Macht, Robert Mamis, Martha Mangelsdorf, Cheryl McManus, Anne Murphy, Kascha Piotrzkowski, Mary Prince, Evelyn Roth, Sarah Schafer, Thea Singer, Jerry

Useem, Edward Welles, David Whitford, and Stephanie Zacharek.

Special thanks to Inc. Online senior editor Leslie Brokaw, the editor of *301 Great Management Ideas,* who suggested the idea for this book and contributed encouragement and ideas along the way. Thanks also to former *Inc. Technology* senior editor David Freedman and to *Inc.* magazine executive editor Jeffrey Seglin. Both of them are ever supportive and excellent sources of all kinds of information.

Finally, this book wouldn't have been possible without the efforts of the group who carefully shepherded it through the production process: *Inc.* Business Resources editorial director Bradford W. Ketchum Jr., senior editor Elyse M. Friedman, copy editor Audra Mulhearn, fact-checker Larisa Badger, and creative director Cynthia Davis of Cambridge Prepress Services, who continues to shape the 301 series design.

—Phaedra Hise
Editor
Boston, Mass.

Editor's note: *Phaedra Hise served as the technology writer for* Inc. *for four years, during which she edited the magazine's "Managing Technology" column and wrote features on the application of technology in small and midsize companies.*

NICHOLAS NEGROPONTE

• INTRODUCTION •

Technology: The Great Equalizer

A young entrepreneur recently went out to buy a car. Her neighborhood Ford dealer pressed her to purchase a Taurus for $19,500. She told him that she needed to sleep on it, but she'd be back the next day. Sleep, however, was not on her agenda. Suspecting that she might not be the only one in her community thinking about buying a mid-priced car, she used the time and the Internet to round up other comparison shoppers. Using e-mail to communicate, she and 15 of her cyberneighbors reached an understanding, and true to her word, the woman returned to the Ford dealer the next morning.

Upon arrival, she informed the salesman that she was ready to buy the car—for $16,500. That price was so far beneath his quote that he hastened to correct her mistake. "No sir, I have made no mistake," she replied. "I simply failed to mention that I am buying 16 cars, not one." Delighted to be selling in volume, the dealer promptly sold the cars at her price.

A buyers' cartel like the one formed by this woman and her fellow Taurus purchasers is almost impossible to create. A cartel requires an adequate collection of participants before it can wield power. Meeting, convening, calling, or just finding enough people with identical shopping lists is enormously difficult, and many consumers don't know how to proceed. They are disadvantaged. Consumers have always been weaker than sellers because they have had no way to shop with equal efficiency. But the marketplace is changing.

Technology has given consumers powerful new tools for finding what they need, what they want, and even what they have never before dreamed of: a new film, a hip restaurant, a timely news article, or a hot

Web site. The concept—collaborative filtering—lets us tap into other buyers' wisdom. It's electronic word of mouth. Collaborative filtering works two ways. It helps us locate the esoteric and difficult to find, and it very quickly blackballs the bull.

We are seeing the beginning of a grassroots kind of *Consumer Reports*—of consumers, by consumers, and for consumers. Successful businesses of every size are those that have acknowledged that not only has technology changed consumer buying habits, but it has also affected the exchange of ideas. Technology—especially its impact on the dissemination of information—is the great equalizer, minimizing the advantages of bigness. Until recently, bigness was a prerequisite to being global. That requirement applied to countries, to companies, and, in a sense, to people. Big nations took care of smaller countries, huge corporations were the multinationals, and the richest were the internationals.

The paradigm is changing, and the change is having a huge effect on the world trade of ideas. In the world of bits and bytes, you can be small and global at the same time. When computing was young, only a few institutions owned thinking tools like linear accelerators. Many people and organizations were in debt to the few who could afford the luxury of science. The debtors were obliged to poach on the basic research provided by those who had the equipment to do it.

Today, almost anyone can own a $2,000 233-MHz Pentium PC with more power than MIT's central computer had when I was a student. Furthermore, with so many peripherals being manufactured and sold at prices every company can afford, nobody is excluded from the multimedia and human-interface arena. This means researchers and other employees of emerging companies can contribute directly to the world's pool of ideas. Being big no longer matters. More than ever before, businesses of every size must trade ideas, not embargo them.

The Internet makes it impossible to exercise scientific isolationism, even if companies want such a policy. We have no choice but to exercise the free trade of ideas. With or without government sanction, the Internet has forced such open exchange, putting the onus on everyone to change attitudes. Small emerging companies can no longer pretend they are too poor to reciprocate with basic, bold, and new ideas.

Before the advent of the Internet, scientists shared knowledge through scholarly journals. More often than not, papers were published a year or more after they had been submitted. Now that ideas are shared almost instantly online, it's even more important that small companies be idea creditors. It's too simple to excuse yourself for being an idea debtor because you lack size or capital.

I have been told that a developing company can only draw from the inventory of ideas that comes from wealthy businesses. Rubbish! In the digital world, there should not be debtor nations or debtor businesses. To think you have nothing to offer is to reject the idea economy. This book is a compilation of ideas from managers of small companies like yours. And like you, they are fighting to make their way in a world where knowledge is the currency. In the new balance of trade of ideas, very small players—even as small as the Taurus purchasing cartel—not only contribute very big ideas, but flourish by doing so.

Nicholas Negroponte
Professor of Media Technology
Massachusetts Institute of Technology
Cambridge, Massachusetts

I

CUSTOMER RELATIONS

"Technology
has pushed
more power to the
front-line workers
in all businesses."

ROBERT B. REICH
Former U.S. Secretary of Labor

• CUSTOMER RELATIONS •

Use Voice Mail to Direct Phone Calls

Because they fear a recorded greeting might alienate their customers, managers at many small companies have been reluctant to install voice-mail phone systems. At the same time, however, most managers do know that voice mail can save both time and money.

First Commonwealth, a dental health maintenance organization in Chicago, receives from 2,500 to 5,000 calls every day. CEO Chris Multhauf says it would be "cruel and unusual punishment" to make a receptionist handle each of them. Besides, he says, "People don't want to talk to an operator. They want to talk to someone who can help them." A **good voice-mail system efficiently routes callers** to that helpful someone.

It took only three and a half years for First Commonwealth to outgrow its first voice-mail system, which cost more than $50,000. But Multhauf was prepared. From the start, he refused to allow the voice-mail system to mistreat customers. By insisting on regular performance reports, which track such details as the number of incoming calls and the amount of time customers are kept on hold, he knew when the original system was approaching the limits of its capacity. Before the old system could become a problem, the company invested some $250,000 in a state-of-the-art system that will easily handle the steadily growing volume of calls.

2
IDEA

COMMUNICATIONS

Write On

For years, White Dog Enterprises, in Philadelphia, had used a computer to produce the White Dog Café's daily menus. But Judy Wicks, owner of the $4.4-million restaurant and crafts store, wasn't happy with the results. "Typeset copy lacks a sense of immediacy. The psychology just isn't right. Specials in your own handwriting communicate fish just caught and lamb just slaughtered. Standard type makes it look as if you've been serving the same hazelnut trout for the past six months."

To overcome that impression, Wicks experimented. She tried to give the menus a just-prepared look by regularly changing the typefaces she used. But new fonts catch on fast, and she noticed her competitors' menus always seemed to look too much like hers.

The solution? **Software that created a typeface font from her handwriting**. "It gives you the best of both worlds," she says. "It's up-to-the-minute style that can't be copied." Wicks has found the software effective for her special menus. It has also given Wicks an efficient way to offer personalized customer service. "I've used it to write letters to customers. People frequently write to say how much they enjoyed an evening at the café, or they might drop me a note to thank me for something, like returning a raincoat." PenFont (Signature Software, 800-925-8840) also makes it easy for her to produce "handwritten" internal employee memos. "It's ironic," Wicks says. "When we first opened the restaurant I wrote the menus by hand because we couldn't afford a computer. Now we've come full circle: We're choosing to spend money on automation that makes us look and feel homegrown."

COMMUNICATIONS

The Fax Will Set You Free

Like many other small-company founders, during his company's early days, Stephen Siegel had to be both executive and customer-service rep. His 12-employee mail-order company, UV Process Supply, in Chicago, sells technical supplies to companies that use ultraviolet light in printing. When customers had equipment problems, they'd call on Siegel and his 22 years of experience in the industry.

It was great for the customers, but Siegel found himself spending about one-third of his time taking calls. Today, the company's Web site handles many customer inquiries, but before the Web was generally accessible, the company found a terrific solution in a **fax-on-demand** program. Siegel estimates that the fax technology, which still fields a significant number of customer questions, quickly reduced his time on the phone by about 85%. Customers calling the fax-on-demand service can choose from hundreds of documents that cover a full range of technical topics. UV Process Supply has integrated the service into its catalog. Each product's description includes instructions for obtaining additional information by fax.

Most of the documents in UV's database are product-information files the company has developed. Those files reside on UV's local area network. Other documents, like product description manuals, are scanned, often by faxing them to a computer equipped with a fax board.

Fax-on-demand services let callers use their telephone keypads to select the documents they want automatically faxed to them. Except for their phone charges, UV's customers pay nothing to use the service.

Laptop Tours: Just Like Being There

Evan Segal is president of Dormont Manufacturing, a $30-million manufacturer of gas-appliance connectors in Export, Pa., near Pittsburgh. Segal wanted to broaden customers' perceptions of what his company could do for them. He knew that if he could get people to visit his factory, he could displace their stereotyped images of Dickensian factories with a realistic vision of his clean, spacious plant and its extensive product line, which ranges from faucets to newfangled coupling devices.

The answer came to Segal as he was taking a virtual-reality (VR) tour of a local university. If he couldn't bring customers to his factory, he reasoned, he should **bring the factory to them**. He decided he would develop a VR tour of his plant and load it onto each of his six managers' laptops. When, however, he discovered that such a project would cost at least $200,000, he quickly changed direction, opting for a multimedia presentation.

Segal hired an independent producer, but he wrote the script himself by pretending he was actually guiding a group around the plant. Now, the company's sales managers pack a 10-minute multimedia tour of Dormont.

The presentation produces real results. One longtime customer took the tour and noticed factory workers assembling faucets—a product she'd never associated with Dormont. It didn't take long for her to file an order for a couple of hundred dollars' worth.

COMMUNICATIONS

Customer Conferences Go Online

Many small and midsize companies doing business on the Internet discover that they can't afford the cost of live product demonstrations and customer training. An inexpensive alternative is to **hold product discussions using Internet "chat" software**, which allows people to communicate in real time.

Companies generally schedule their event-based chat sessions in advance, so they can alert customers, colleagues, and the press. At the designated time, participants log on to the Web site and click on a chat-room icon. The chat program can also be set up so that any visitor wandering through the site can request a session. Clicking on an icon, visitors notify a designated company employee that they want to participate.

SenseNet, in New York City, used chat software to confer with a client about an ad campaign. "We had people scattered all over the country, and so did they," says Art Thompson, chief technology officer. "We could have patched together a conference call, but that would have been harder to organize and a lot more expensive." Using chat, the group "discussed" the campaign in one window while viewing the artwork in a second window.

"A lot of us are not good note-takers," says Thompson. "With this, you get a transcript that you can store, search, and e-mail to people who could not make the meeting."

Welcome to My Web

When Cathey Cotten, owner of MetaSearch, a $360,000 high-tech employee-recruiting firm in San Francisco, wants to forward information about potential job candidates to her clients, she no longer bothers with faxes or letters. Instead, Cotten posts résumés at **password-protected customer-only Web pages** that are part of her firm's Hire Site. "The turnaround time is great," says Cotten. "If you fax a résumé, it can sit there for days or get lost." Cotten's system alerts her when a client reads a résumé, schedules interview times with candidates, and reminds her to follow up with a call.

Doug Wright, president of $2-million Wright Communications, in New York City, also uses password-protected Web pages. The graphic-design company's site lets clients view work in progress, approve changes, or sign off on jobs by e-mail. And because he can easily set up a new client page in less than half an hour, Wright also uses the site for selling. He prepares a page for each prospective customer even before he's made the first sales call. "It makes us look like one of the big guys," he says.

COMMUNICATIONS

Voice Mail: Not for the Customers

Slapping a voice-mail system on your phone line can save your company time and money, or it can irrevocably upset callers. The solution may be to have an automated attendant greet one group of callers while retaining the services of live operators for those who really appreciate the human touch.

Tami Simon is president of Sounds True, in Louisville, Colo. Simon resigned herself to voice mail for her $6-million mail-order seller of spiritual audiotapes after she realized her telephone operators were spending way too much time taking messages for their coworkers. But Sounds True doesn't treat all callers the same. Customers who dial Sounds True's toll-free number still get a live operator, and those who phone the company's corporate number hear the voice of an automated attendant. Simon says the $7,000 voice-mail system, which the company bought from its telephone vendor, is a big success. Nevertheless, with no one screening her calls, she finds that picking up her phone is "a little bit like roulette."

Levenger, a $65-million mail-order seller of reading accessories, also has flesh-and-blood operators answering customers' calls. Company president Steven Leveen says, **"We don't subject our customers to voice mail, only our suppliers and ourselves."** The company, in Delray Beach, Fla., paid about $30,000 for its sophisticated voice-mail system. A special feature allows Levenger personnel to screen calls. Leveen appreciates the computerized voice that requests each caller to state his or her name. Only after he hears who is calling does he decide whether or not to answer his phone.

Too Many E-mail Accounts?

When Dana Free was advertising coordinator for OGCI Training, a 25-person company in Tulsa, she found herself managing several e-mail accounts and services. The company, which offers training services to petroleum companies, had added its second e-mail service because it offered different features that proved irresistible. Rather than track down ever-changing e-mail addresses and send out change-of-address announcements to all her contacts, Free decided to keep both accounts. After all, in spite of the features that had attracted her to the second service, she still found the first service more reliable when it came to exchanging large attached files with clients.

Free could have set herself up for e-mail hell, but she discovered a solution that can **manage several e-mail services**. Free configured E-Mail Connection (ConnectSoft, 800-889-3499) to check each of her services. When she logged into E-Mail Connection, she would see a single list of her messages from every service, and she was able to open all of them in the same window. Not leaving anything to chance, she used the Default Destination setting to specify the e-mail address to which she wanted each recipient to reply. "With all the attachments going back and forth," Free says, "I wanted to be sure that people sent them to the right service."

COMMUNICATIONS

Check Your Service: Call Your Company

What's the best way to make sure your service procedures don't abuse your customers? Take the advice of William Pape, cofounder of VeriFone, an electronic-payment processing company in Redwood City, Calif. He recommends finding out what it's like to **be your own customer**. See what happens when you call your service numbers yourself. If you're afraid one of your employees will recognize your voice, have a friend make a call while you monitor the entire transaction. As your company's customer you'll quickly discover whether, for example, your voice-mail menu is too complex or extensive. And if you have wondered whether people are kept on hold for unreasonably long waits, this is the best way to find out.

To introduce real change, arrange for senior managers to experience customer service at its lowest acceptable point. For example, if company standards allow customers to be left on hold for as long as 10 minutes, make sure your managers experience a 10-minute hold. Also, give managers the opportunity to staff your customer-service lines. That immediate exposure to your customers and their problems will quickly familiarize management with the obstacles your employees confront when they try to meet customers' expectations. VeriFone has grown far beyond the start-up stage, but it still applies the same procedures to monitor and improve its customer service.

COMMUNICATIONS

How to Make Your Phone Customer-Friendly

Here are some ways to make sure that your phone system doesn't make your callers feel as if they've landed in **voice-mail jail**:

- Design your system to inform customers how long they can expect to spend on hold. They will appreciate the opportunity to decide whether to wait. And give callers choices. For example, "Bob is out of the office right now. If this is an emergency, press 1 and you'll be connected to his cellular phone. Otherwise, press 2 for voice mail."
- Your voice-mail system should regularly generate simple one-page performance reports that answer such questions as: How long does it take callers to reach a human? What percentage of callers hang up before connecting with the appropriate person? How many transfers do callers endure before they speak with someone who actually handles their problems? What percentage of callers have their problems resolved? How long does it take to get a satisfactory response? If you follow those benchmarks, your company's voice mail can be a helpful customer-service tool rather than an annoying barrier to customer satisfaction.

COMMUNICATIONS

Go Global—Electronically

Communication with customers is a source of competitive edge. You can improve your products and services if you learn from your markets. Before it was acquired, Patrick McDonnell was CEO of a small manufacturing company in Andover, Mass. He describes how his company benefited from staying in touch—electronically.

"Information technology lets small businesses **communicate with customers and suppliers without regard to time, place, and sometimes even cost**. Historically, that capability was limited to large companies because only they could afford to have worldwide marketing teams. Even when the company had only 20 employees, e-mail allowed us to communicate with customers all over the world—without having to travel. That's important because we were selling one of our products in Europe and Asia, and one of our strategic partners has offices in Munich and Singapore.

"Using teleconferencing, we held an international sales meeting with our Canadian partner and gathered marketing information from customers around the globe. The data told us we needed to add performance capabilities and modify some features in our emissions-monitoring instruments—features we had considered unimportant. With the help of the Internet and e-mail, we were also able to form stronger partnerships. And with strategic partnerships, small businesses can develop vertically integrated teams—like ours with the Canadian company—that have the flexibility and versatility necessary to compete with large businesses."

Trimming Time at Telephone Tag

Like any other good tool, voice mail works well only when it's used intelligently. R. Douglas Shute, who at the time was CEO of Steinbrecher Corp., in Burlington, Mass., explains that there is no point in using voice mail if you treat it like an audio While-You-Were-Out pad. Since he learned how to take advantage of the particular strengths of voice mail, he uses it much like a reliable administrative assistant.

"Although we have many new modes of communication, we have not exploited them fully in terms of connecting with our customers. Take voice mail. Much of the inflection and emphasis of face-to-face interchange can be communicated by voice mail. Messages can be saved, passed along, returned, or dumped—individually. Responses are equally flexible. But how many times have you left a customer a voice-mail message saying 'This is so and so. Please call me back at…'? If instead you **rattled off a quick question and then asked for a voice-mail answer**, you'd get more responses and avoid wasting customers' time on telephone-tag calls."

COMMUNICATIONS

Meet Customers Online

It was just another day on the job at Invitrogen, in Carlsbad, Calif. One of the company's engineers was catching up on his reading. He'd read Internet postings about the sex life of yeast (bionet.molebio.yeast), and then he'd happened upon a newsgroup that focused on DNA cloning products like those made by his employer. When he saw that a user was trashing one of Invitrogen's products, the engineer fired off a prompt response. He identified himself as an Invitrogen employee and offered tips on how to make better use of the product.

The engineer's message was the gut reaction of one scientist offering to help another. He didn't realize that he was taking a first step toward establishing an online presence: **interacting with customers online**. Of the Internet's thousands of newsgroups and proliferating e-mail mailing lists, many are devoted to discussions of specific industries, products, or services. When customers, suppliers, and competitors mix it up online, there are plenty of opportunities for controlling damage, providing service and support, and collecting feedback.

These days, Invitrogen, with annual sales of $28 million, assigns a full-time company representative to scan newsgroups and mailing lists, looking for messages about the company and its products. That person also handles all technical help questions sent by e-mail. "Before," the engineer asks, "who cared if someone in North Dakota had a problem? Who was he going to tell? The Internet has changed all that."

Leave Town, but Stay in Touch

George Matarazzo wanted to get away from the city—not from his customers. So he closed down Matarazzo Design, his high-growth, high-profile landscape-architecture business, and started a new business, Matarazzo Land Planning Consultants, in a barn on his 80-acre farm in tiny Wilmot Flat, N.H. He did have one problem, though. Some of Matarazzo's old customers couldn't find him.

Matarazzo's business depends on contacts and a nationwide reputation he has built over many years. People who had worked with him in the past were trying to reach him in Concord, N.H., the home of his previous company. The few persistent souls who did finally reach him in Wilmot Flat told him that they had been trying for months to locate him.

Matarazzo found a simple solution. For a modest monthly fee, the telephone company lists numbers for Matarazzo Land Planning in both Concord and Wilmot Flat. Anyone who calls directory assistance in Concord gets the Concord number, which **automatically transfers the call** to the Wilmot Flat line. Matarazzo says that even though he racks up the long-distance charges for those transfers, the service never costs him more than $25 a month. And, for a business in which a single call from an old client can yield months of work, that's a price Matarazzo is more than willing to pay.

COMMUNICATIONS

Harness the Power of the Net

Net Gain: Expanding Markets through Virtual Communities, by John Hagel III and Arthur G. Armstrong (Harvard Business School Press, 800-338-3987, 1997, $24.95), speaks to the authors' belief that, thanks to the Net, we're in the midst of a radical transformation: Buyers have wrested power from sellers. To **identify, understand, and gain the loyalty of "virtual communities" of customers**, you'll need to invoke new strategies.

"Many companies new to the Internet use the same formulas they depend on in the physical world," Hagel explains. "That may allow them to do things faster, but they miss real opportunities for rethinking how they do business and redefining who their customers are. They should look at going on the Internet as they would look at going into a new country—and exploit the potential unique to the new place.

"The Motley Fool Financial Forum, an investment group on America Online, has done that well. The two brothers who started the group originally wanted to do a newsletter. But they found, on AOL, that the real power lay in drawing people together to interact and share information. They created an engaging environment by allowing people to discuss all kinds of investment strategies, rather than expecting them to read a static newsletter online."

CUSTOMER TRACKING

Prevent Information Backlash

Marketing experts trumpet the benefits of following customers. Don't forget, however, that a little low-tech sensitivity to customer privacy also has its place.

Put yourself in the position of a customer: Imagine yourself stopping at a store and mentioning to the friendly salesperson that you're looking for a gift for a special someone. You are unaware that the details of your purchase will be typed into the shop's database as soon as you leave the store. Months later, long after your "special" relationship has soured, you return to the store. How will you feel when the salesperson, a complete stranger, eagerly starts making suggestions for romantic purchases, based on your earlier visit? Many customers get the creeps even thinking about a store's tracking such personal information as their birthdays or clothing sizes.

So what's a retailer to do? The solution isn't rocket science, says Stephen Silverman of Silverman's, a men's apparel chain in Grand Forks, N. Dak. It's just a variation of the one-to-one marketing concept. Be sensitive enough to **recognize the customers who are offended by information gathering**. Train your salespeople to put customers at ease before asking for their addresses. Make sure salespeople recognize signs of reticence and know when to back off. Nobody should wave even marginally intimate knowledge in the face of an unfamiliar customer. And what do you do about customers who hate being tracked in any way? That's what you need to note in their records.

CUSTOMER TRACKING

What Do Your Customers Really Think?

Gary Hirshberg, president of yogurt maker Stonyfield Farm, in Londonderry, N.H., was confident that the company's apricot-mango flavor would take off, but even he wasn't prepared for the accolade it received from one customer. So taken was the woman with the new product's color that she painted her bedroom to match it.

Hirshberg can tell you many other stories, because Stonyfield Farm solicits and tracks its customers' opinions. Thanks in large part to a friendly note on every yogurt container, the company gets about 150 customer calls and letters each week. The gist of **every customer message is entered into a database**, and Hirshberg and other company managers mine it for opportunities to cement customer loyalty, explore promising new niches, and fine-tune the product line.

For each customer-service call, a rep completes an electronic fill-in-the-blanks "form" that asks for the customer's name, address, phone number, his or her complaint, comment, or suggestion, and Stonyfield's response. The company can sort the information in any number of ways—by region, time period, type of complaint, and so forth.

Stonyfield Farm rarely makes a product decision that has not been influenced by the information stored in that database.

Tame Your Data Beast

Is your database under control? Consider these tips from the experts:

- *Be choosy about the information you track.* Such minutiae as fraternity membership and each customer's favored color won't pay off. Maintenance on a sprawling database is a bear.
- *Develop a simple rating system.* As you enter more information, it should become easier for you and your employees to **rate each customer's importance to your company's growth**. Based on such variables as sales, future sales, and referrals, assign each customer a score of one, two, or three. Set up a similar system for ranking leads.
- *Keep your list clean.* Give up-front thought to how you want your staff to enter names. Don't leave such choices up to your clerks. If you don't take that precaution, a customer like IBM might also appear as International Business Machines. Use contact managers' drop-down menus or pop-up boxes that prompt users to fill in fields correctly. A temp entering trade-show leads can quickly corrupt a whole database.
- *Identify the best leads, and turn them over as quickly as possible.* Put questionable prospects into a "holding" database, and move dead leads to a purge file. The chief objective should be strength, not length.
- *Be realistic.* Unless your people regularly enter up-to-date information in your database, it won't be much use. One company with computer-shy execs has them relay information about customers they've visited to a dedicated voice-mail box. An assistant transcribes the messages and enters them into the marketing database for all to see.

CUSTOMER TRACKING

Focus Your Data

What do you track with your sales database? Today's database software is overwhelmingly powerful, giving us the ability to track too much information. It's a real trick simply to **focus our databases on what we *need* to know**. Kim Whittaker, president of Baby Faire, in Winchester, Mass., puts on consumer shows for new parents. Her company, with four employees and $650,000 in sales, relies on ACT! 2.0 (Symantec, 800-441-7234) to follow 6,800 customers and leads.

"We track objections or why someone said no, we note whether a prospect is local or national, and we include a one-sentence business description," Whittaker explains. "I also ask customers what we should do differently next year. In the notes field, I keep all sorts of miscellaneous information and do keyword searches. I track news on companies so I have reasons to call back. Because of what I do, I also enter information on customers' kids."

Charles Forbes, principal of Earnings Performance Group, in Short Hills, N.J., consults to banks. The 100-employee company uses GoldMine 3.2 (GoldMine Software, 800-654-3526) to keep information on some 3,000 customers and leads. "We use the customer history file extensively. It gives us the anatomy of a sale and a running dialogue—up to two years long—on all appointments, letters, and proposals. We also use the database to measure sales productivity. We can check, for example, a salesperson's calling pattern. With 10 sales reps around the country, the remote-synchronization feature is really important for us. We get weekly updates on all sales activities, and we can see what our reps' calendars look like."

CUSTOMER TRACKING

Use the Net to Keep Reps Informed

As an ever increasing number of its clients opted for e-mail rather than telephone communications, talent agency Wolgemuth & Hyatt, in Brentwood, Tenn., needed to find an efficient way to monitor its correspondence. Four years ago, the $1-million agency traded in its old contact-management software for a system that came Internet-ready. The new system's e-mail feature and its ability to automate such functions as contract writing helped the firm double revenues within a year.

"With all the details taken care of, we were able to handle twice as many customers," cofounder Mike Hyatt explains. The new system **automatically logs each e-mail message into the appropriate customer file** of the database. Wolgemuth & Hyatt sales reps can easily bring themselves up to date by getting that information through the Internet rather than phoning in for messages. And, Hyatt notes, going onto the Internet usually costs less than long distance.

CUSTOMER TRACKING

Control Your Need to Know

Since 1990, Ann Cavoukian, coauthor with Don Tapscott of *Who Knows: Safeguarding Your Privacy in a Networked World* (McGraw-Hill, 800-338-3987, 1997, $24.95), has served as assistant commissioner and commissioner of the Information and Privacy Commission for the Canadian province of Ontario. She argues that it makes good business sense for companies to respect customer privacy:

"Businesspeople traditionally rail against the notion of privacy legislation, claiming that it impedes free enterprise. But that's not necessarily true. When Quebec recently extended its privacy laws to the private sector, businesses were not crippled, as many feared. In fact, privacy codes may actually help you gain customer trust and loyalty. Some companies have even found that **privacy protection is a cost-reduction tool**. Companies often have archaic information practices, and they collect a good deal of information from their customers—information they just don't need. A company that begins to scrutinize its information holdings from a privacy perspective may discover that it can save valuable computer processing time and memory. And it might also find that its employees will be more efficient if they don't have to collect data that never gets used."

CUSTOMER TRACKING

Party On, with Instant Info

Rita Bloom, president of Creative Parties, in Bethesda, Md., now runs her entire business using groupware. To **keep close tabs on her clients**, the owner of the $3.5-million events-management company relies on Lotus Notes (Lotus Development Corp., 800-343-5414), for which she paid a hefty $45,000.

From their workstations, staff members can retrieve information about every phase of the event-planning process, in just about any form. When Bloom enters a client's name, she can view correspondence, floor plans, schedules of events, invoices, even scanned-in photos of fabric swatches and musicians. And if a client asks her to add, say, the name of a new master of ceremonies, Bloom quickly calls up a new vendor sheet and fills in the necessary information.

It takes only seconds for any of Bloom's employees to respond to urgent questions from clients and suppliers. For example, a caterer once called with a desperate last-minute query: Should the tablecloths be pure white or yellow and white? Bloom turned to her computer and located a picture of the fabric swatch for guests' tables. She could see that the mustard-yellow floral print would likely clash with the yellow and white cloths, and she instructed the caterer to use the pure white.

Since she installed the groupware, Bloom has expanded her business beyond nonprofit organizations and private individuals to corporations. "We don't look like a small-town mom-and-pop operation anymore," she says. "We look like a 21st-century company."

CUSTOMER TRACKING

Ask What Your Database Can Do for You

To design a customer-service database that works for you, you need to answer some fundamental questions. What kind of customer support will you offer? Do you have sufficient staff to answer the phones? What kind of information will your phone reps need to respond to customers' questions? Where is that information currently stored? Beyond answers to these basic questions, you'll need to consider these other issues as well:

- *The electronic future.* Make sure any new technology you invest in will work with the computers and databases you already have. While you may not be ready to link all your computers today, in the long run, it still makes sense to **have information go directly into your computer**.
- *The long-range plan.* Your immediate need is to use technology to help you answer your customers' questions. The ultimate goal is to mine their comments for business intelligence you can analyze and use to grow your company.
- *The low-tech touch.* Remember that, even today, many of the best techniques for making your customers happy are low-tech and old-fashioned. Used wisely, technology can help you offer that personal touch. Don't go so high-tech that your customers feel alienated.

CUSTOMER TRACKING

Why It Pays to Keep Data Current

One-to-one relationship selling is the oldest game around, but if you're not working it in conjunction with a well-maintained customer database, you're letting opportunities slip away.

Silverman's, a men's-apparel chain in Grand Forks, N. Dak., has built close relationships with many of its customers, and its database is filled with up-to-date information on individual shoppers' sizes, buying habits, and preferences—even details about items they tried on but didn't buy. Salespeople use the information to help people who are buying gifts and to follow up on big sales. Two weeks after every sale, the computer automatically generates a report that reminds the salesperson to **call customers and confirm that all purchases were satisfactory**. Those calls are crucial to preserving good customer relations. "Most customers would rather stop shopping at a store than take time to complain," says third-generation clothier Stephen M. Silverman.

The store's marketing department uses the database to target its efforts. If, for instance, the company wants to promote the arrival of a designer's new line of winter clothing, it's a simple matter for the database to produce a mailing list of customers who had purchased items from that designer's warm-weather line. Silverman says that within four weeks, such mailings typically elicit 25% response rates. That's without the lure of a discount.

SERVICE

Better Info Access Means Better Service

Garland Heating and Air Conditioning Co., in Garland, Tex., which specializes in the installation and repair of heat, ventilation, and air-conditioning systems, took quite a while to upgrade from paper to computer systems. When customers started to complain, service coordinator Rick Kelley realized that the company had waited too long.

Before the $5-million company installed its electronic database, a customer calling for repair service had to wait while Kelley sorted through filing cabinets for a record of that customer's repair history. Once he found the record, he had to skim through it to find the name of the repair person who had last visited the site. Finally, he had to scan a handwritten schedule to see when that service person was next available. "The paperwork we had to search through was unbelievable, and the search was very time-consuming," recalls Kelley. "We just couldn't react that fast. Customers grew impatient with us."

Now when a customer calls to complain about a ventilation system that has gone on the fritz, Kelley can enter the company's name into the database and retrieve the **company's service-repair history in an instant**.

When emergencies arise, Kelley really appreciates having immediate access to the information he needs. For example, the frantic manager of a small supermarket calls because a broken meat freezer is threatening to let $100,000 worth of meat go bad. Kelley consults the computer, pages the appropriate technician with a special signal, and sends him to the grocery store with all the information necessary to make an efficient service call. "We've improved our efficiency by 90%," says Kelley.

SERVICE

LAN Preserves the Personal Touch

A computer network can elevate your small company to a higher level, endowing it with the sophisticated information-management capabilities of the big guys while sidestepping the bureaucracy that comes with size. With a sales staff of 25, Sheila Skolnick, owner of Elite Cos., an $18-million hotel-supply business in Setauket, N.Y., prides herself on the service Elite delivers to its 4,200 customers.

Skolnick, who started the business in 1984 in a spare bedroom of her home, says, **"The network makes us look like a billion-dollar company."** She credits her 12-node local area network with allowing her to triple the size of her business without adding to her payroll.

The network was Skolnick's response to a paradox that had troubled her since the early 1990s. "I was afraid that if we grew too big, the company would choke and die," she says. Skolnick had built the business by giving individual attention to her customers, going "back of the house" to talk with general managers, and riding shotgun with housekeepers as they scrubbed bathrooms and made beds. But as her customer and supplier lists swelled, the company began to show unmistakable signs of organizational breakdown. While she was out rounding up new accounts, her crew back at the office could barely service those she already had. Instead of cross-selling, salespeople were struggling to find files, invoices, and manufacturers' product books.

"We needed to function as one big brain," says Skolnick. The network let her company grow while maintaining personalized service.

SERVICE

Instant Maps Speed Deliveries

If your customers' ardor has cooled so much between the time they order and the time you can deliver that they refuse to accept your product, shouldn't you find a way to provide next-day service?

The flood of sales lost due to the long wait for delivery had dampened the spirits of management at Sparkling Spring Water, in Highland Park, Ill. The $15-million company installs coolers and delivers bottled water in Illinois, Indiana, and Wisconsin. The company was losing not only customers, but also the time spent in futile attempts to deliver the cooler.

All that has changed. The company now uses **mapping software to create daily delivery routes**. When new customers order the service, the reps find out when and where to deliver the cooler. That information goes directly into a database on the company's network, and a supervisor transfers the information to MapInfo (800-327-8627) mapping software. When the supervisors are ready to build the next day's delivery schedule, the computer combines the customer information with the number of drivers available and generates the route each of those drivers will follow.

Within one month of installing MapInfo, the company's order cancellation rate had fallen from 32% to 27%, and it has since dropped to 20%.

SERVICE

Let Customers Track Their Packages

If your customers have Internet access, you and they can take advantage of the Web sites that UPS and FedEx have set up. Your customers are probably accustomed to having you trace the packages you send them, but if you give them the shipping numbers, your **Internet-ready customers can trace their own packages** on the shipper's Web site. Your customers will appreciate having a sense of control over their deliveries, and your reps will be free to handle a larger volume of calls.

That process has worked well for FMC Resource Management, in Monroe, Wash. The independent subsidiary of Merrill Corp. distributes printed promotional materials. If an impatient customer is frantic to know when a parcel will arrive, he or she simply calls an FMC customer-service rep for the package tracking number. The customer then connects to the Internet, goes to the UPS or FedEx site, and, in an instant, knows when to expect the order. "We wanted technology that would get our customers the information they need, when they need it," says CEO Mark Trumper.

SERVICE

Net Nets Market Intelligence

Consultants at Balentine & Co., a midsize investment firm in Atlanta, **turn to the Internet for hard data**. Gary Martin and the firm's other consultants bolster their advice with information they find on the Net.

Not long ago, during a lunch meeting, one of Martin's clients mentioned his plan to develop a design for a plastic wheelchair. But, he said, he wasn't sure how to gauge market interest. Martin offered to browse the Net to see whether he could turn up any leads.

Turn up leads, he did. Using a few keywords, Martin quickly found a newsgroup for people with disabilities. He posted a message asking users for feedback on his client's idea. Within a few weeks, Martin had heard from enough wheelchair users to be able to pass along encouraging news to his client. Many respondents related that they owned three wheelchairs they kept in different locations, and they found the prospect of one lightweight portable chair very appealing. Today, Balentine & Co. includes intelligence gathering on the Internet as part of its basic menu of services.

SERVICE

Voice Mail Tops Telemarketers

Anticipating a good response to his television ad, Menderes Akdag of Lens Express, in Deerfield Beach, Fla., hired a telemarketing firm to field calls for brochures. The response to the spot for the company's replacement contact lenses was, however, so enthusiastic that nearly 40% of the callers got busy signals.

Akdag, president of the $48-million company, decided it was time to see whether a **voice-mail system did better than the telemarketers**. Callers to the toll-free number featured in the ad now hear a menu that invites them to press 1 for a free brochure or 2 to order from an operator. Because 85% of the callers want the brochure, they never need to speak with an operator. The automated system easily handles thousands of calls at once, and Akdag is no longer troubled with busy-signal problems. What's more, he's cut his costs by some 50%. The call volume that once accrued to the telemarketing firm now belongs to Lens Express, and the increase in volume has qualified the company for lower phone rates.

SERVICE

Give Old PCs to Your Customers

Karen Goode calls it her computer "graveyard." Like many small-business owners, Goode recently took her company through a major computer upgrade that left her with a collection of old 386 PCs. But she found she could put some of the old computers back to work. Her solution saves time for her customers and money for her business.

The 18 employees at Goode & Associates, in Berea, Ohio, turn recordings of hospital and clinic dictation into printed text. Because clients want the transcripts as soon as possible, it wouldn't take much time at all to rack up extensive bills for overnight mail and courier service.

Goode decided to **offer her customers the leftover 386s** so that she could transmit their transcripts by modem. Four of the smaller accounts accepted, and Goode bought communications software for those who needed it. After educating herself about the transmission process and taking a short course that familiarized her with the word-processing software many of her clients use, Goode was ready to help her customers and staff understand the new procedure.

Today, the use of modems frees up Goode's printers, saves postage, and satisfies customers' passion for fast turnaround. "They're thrilled because they get it the same day it's dictated," Goode says. "I don't have to pay for postage, and I don't have to pay a clerk to print and mail." She estimates that for each account, she saves $60 to $70 in weekly overnight-mail costs. Once Goode could demonstrate how well her new system was working, several of her large customers agreed to switch to modem transmission. But they supplied their own hardware.

II

"The Internet
is an incredibly cheap
and far-reaching
distribution channel.
All a small company
has to do is be good,
and the world will beat
a path to its door.
That's never been true before."

DAVID BIRCH
founder and president of Cognetics Inc.,
Cambridge, Mass.

ADVERTISING

Business Cards on a Disk

If you ask Seth Resnick for his business card, the Boston-based photographer and video producer will hand you a diskette. When you stick it into a Mac and click on its icon, you're treated to a 30-second video highlighting Resnick's work. In fact, lots of companies have thought about passing out commercials-on-a-disk, but they've been frustrated by a technical barrier: To get 30 seconds of decent-quality video, you normally need at least eight megabytes of disk space. That's more than four times the capacity of a diskette.

So how did Resnick manage? He used a technique that squeezes the video, throwing out image elements that aren't crucial to the quality. When he saw how well the technique worked, he started showing his digital business card to other businesspeople who might want to follow suit.

Now an auto manufacturer and a major software vendor are among those negotiating to license the scheme and produce their own video diskettes. They aim to **mail out diskette-based commercials** to a target audience for far less than the cost of a television ad or a multimedia presentation.

"CDs and multimedia are overkill," Resnick says. "For something like this, simplicity is key."

ADVERTISING

Advertise Your Web Address

John Honiotes, director of new business for Auto-By-Tel (ABT), in Irvine, Calif., says that the ranks of salespeople will shrink when most markets respond only to the computer-literate, softer sell. "It's a classic case of survival of the fittest," he says. "Many will embrace the new technology. Others won't and will have to find something else to do."

ABT founder Pete Ellis warns that the balance of power is shifting into consumer hands, and companies will need different strategies for attracting customers. Technology, Ellis understands, has affected automobile sales—his industry. For 50 years, the true cost of a car was a closely held secret that manufacturers and dealers kept to themselves. Now, through the Internet, buyers have access to that information—24 hours a day. Consumers have the edge, and dealers who can't deliver low, haggle-free prices and superior service will not survive. "Car dealers don't like being told how to run their business, but in a couple of years a lot of these guys aren't even going to be in business," says Ellis. "The biggest thing they've got to figure out is how to drop their costs. I see the **Internet destroying the old structure**."

When his now $30-million company had sales of just $5 million, Ellis decided to spend $1.2 million on a Super Bowl commercial. ABT's bold move paid off. The company became known as the first Internet-based company to advertise on the Super Bowl.

ADVERTISING

Frugal Move: An Online Catalog

Drew Munster had meager financing when he started Tennis Warehouse, a direct-mail business in San Luis Obispo, Calif. He could afford only a black-and-white, two-page catalog and almost no photos. How did he compensate? He created a Web site and put his mailer online. And he **kept printing and postage costs to a minimum while expanding his catalog**.

The old two-page catalog, which cost $1,200 to print and mail every two months, went to a home-grown list of 3,000 names, and it listed only half of Tennis Warehouse's line.

The Web version, a 30-page electronic catalog listing all 200 of his products, took Munster just two weeks to produce. About 50 items are linked to color photographs and text that details product features. It's a simple matter for Munster to update his inventory, ad copy, and prices at a moment's notice.

In his first six months online, Munster saved $1,000 in printing and postage. Local Internet service was costing Munster $30 a month, and he reports that his largest expense has been the $60 a week he pays to the clerk who responds to e-mail requests for additional product information.

He credits the site with at least 25% of his company's growth. "We have all the business we can handle," Munster says.

ADVERTISING

Hits That Rate Attention

When Windham Hill Records, in Beverly Hills, Calif., set up its Web site, its marketing people made sure they would be able to track its popularity. "Just as with a print ad, you want to know the audit numbers to compare the cost per thousand," says Kristen Schlesinger, associate manager of strategic marketing.

But many Web-service providers offer "hit rate" information that fails to report the actual number of visitors. For example, if five people each make three quick visits, they register as 15 "hits." Likewise, one person looking at a page with six graphics registers as six "hits." Windham Hill insisted that its Web service provider deliver **legitimate detail about visits to the Web site**, specifically requesting daily user-feedback reports tracking the site's more than 1,000 daily visitors.

"On the Web, we know what ads people look at, for how long, and how involved they get. We can break the data down by sections—each recording artist's page," says Schlesinger. "That tells us who's more popular." Windham Hill trusts those detailed reports enough to make daily changes based on them. For example, when the data indicated that artists whose names started with A and B were significantly more popular than others, the marketing department rearranged the list, from alphabetical to an order that emphasized other artists.

ADVERTISING

Traditional Approaches Lure Customers to the Web

The electronic way to promote your Web site is to list it with such search engines as Yahoo!. Those giant indices are where Web searchers use keywords to find specific information. But many companies make the mistake of relying too heavily on search engines to bring customers to their Web sites. Alan Klotz of PhotoCollect, a small photography gallery in New York City, found that **print promotion worked better**.

After his site's debut, Klotz registered it with all the search engines, following up with visits to test his listings. He was horrified to discover that none of the keywords he tried placed his Web site at the top of the list. After having spent more than $7,000 on his site, Klotz was anxious indeed.

Turning from the electronic to the tried and true, Klotz advertised his Web site in the Sunday *New York Times* and wrote a press release, which he sent to industry magazines. The press release worked. Two magazines later covered his site in feature articles. "Traffic at the site increased, and people referred to the magazine articles," he says. The $125 newspaper ad attracted attention, too. Within days, he reports, "I got eight calls saying 'What are all those letters above your phone number?' Good thing I ran that phone number."

Klotz makes his high-ticket sales over the phone or in person, and when customers call the number listed at the site, Klotz asks where they first learned it. "About a third say through the *New York Times* ad," he says. "The other two-thirds are from the magazine articles. Nobody has ever said they found me through a search engine."

ADVERTISING

Stylish Web Sites

"In fashion," says Kerry Clark, formerly an assistant designer at upscale clothing maker Nicole Miller, in New York City, "the eye is always growing tired. What it sees one season, it doesn't want to see the next." The same goes for Web sites, Clark says, only the "season" is shorter.

Management realized how true that was only after having paid a design firm $10,000 to build a Web site that company founder Nicole Miller later characterized as "crap." She quickly decided that the company should oversee the site's radical redesign.

The redesigned version, financed from the company's $1.2-million advertising budget, featured such interactive icons as TV screens with video clips of models dancing down a runway to hip music.

Clark objected to characterizing the change as a "makeover." That, she said, "implies a before and after rather than a continual evolution and growth with technology. A **Web-site redesign is never really over**. You never really figure it out. Just when you think you've created the coolest site in the world, you find out there's something even cooler."

ADVERTISING

Investments in Customer Confidence

In addition to its World Wide Web site, New York City-based Kaplan Educational Centers maintains sites on America Online and the Microsoft Network. Chief executive Jonathan Grayer headed up the push to go online in 1994, establishing sites that direct-market, elicit customer feedback on test formats and new products, and distribute test-preparation software and campus guides on CD-ROM. Despite the diversity of Kaplan's offerings, Grayer says the online ventures of the test-preparation enterprise have yet to show a profit.

Why does the company bother to make the investment in so many online sites? "We can't afford to be behind the curve. Our customers are buying confidence—the ability to walk into a test feeling good. If we're in the business of selling confidence, part of building that confidence is having **customers feel that we keep them out in front** in educational and technological developments."

ADVERTISING

Direct E-mail Club

For Big Emma's, a Boston company that sells secondhand movie laser discs, direct-mail promotions meant both high costs—each mailing cost $4,000—and extended lag time between marketing and sales. Despite fevered demand for the product, discs were sitting in a warehouse for six weeks before they were sold. It took the staff, which at the time numbered only two, nearly a month to produce inventory lists. By the time customers placed their orders, those lists were already out of date. Customers were frequently disappointed, and Big Emma's was barely breaking even.

The solution seemed obvious: ditch traditional mailings and go online. But that raised the problem of how to get online customers to the site where they could check out the up-to-date inventory lists.

Cofounders Christian Strain and Liam Sullivan set up a Web site on a virtual server run by a local Internet service provider (ISP). Then, to get the word out, Strain turned to a newsgroup dedicated to laser discs. After two weeks of simply following the group's postings, he posted his own message announcing Big Emma's **new e-mail club**.

Membership is free, club members receive Big Emma's list of titles before other customers, and there's no obligation to buy a thing. Membership in Big Emma's e-mail club has exceeded 3,000, and the organization costs virtually nothing to run. The basic ISP fee is just $75 a month. Those who aren't members can see the inventory list at the company's Web site, where anyone can place orders. Big Emma's revenues shot up 400% in just one year.

ADVERTISING

Web Promotions 101

The market for online erotica is exploding, partly because purveyors of erotica are smart about **how they advertise their Web sites**. Here are seven lessons from the industry:

1. Redesign your site regularly. If your site doesn't change, people assume you have nothing new to offer.
2. Continually upgrade to ever-friendlier user interfaces, and use the latest generation of multimedia technology.
3. To expedite service, be your own access provider. Speed is crucial.
4. Promote yourself in newsgroups that cater to your industry.
5. Distribute come-ons and giveaways that demonstrate the range of your services.
6. Be innovative. Unique content matters less than innovative organization, presentation, and information management. It's worth sending your customers periodic e-mail deliveries that advertise search-engine updates.
7. Design your site for the international market. After all, the rest of the world is targeting *your* customers.

ADVERTISING

Climb onto the Web, Painlessly

An electronic mailing list delivers information around the world instantaneously, and its cost is a tiny fraction of paper direct mail. It's also an effective way to establish online customer loyalty before you invest capital in a Web site. Mountain Travel-Sobek, in El Cerrito, Calif., chose that conservative route to the Web.

The adventure-travel company started cautiously, first e-mailing a weekly newsletter. The three- to four-page letter, "Hot News," reported recent news on travel interests like the current status of Ebola in Zaire and Katmandu's political climate. Content for the weekly newsletter comes from the company's offices and guides all over the world, and company personnel compile issues from those bulletins, descriptions of upcoming trips, and notations of available discounts.

After the travel company, which employs 35, had amassed a list of 1,500 subscribers, it upgraded its electronic offering to a full-fledged Web site. By making good use of its subscriber list, it **guaranteed Web-site hits from the very start**. Now, 77,000 people visit its site every month, and business continues to grow.

ADVERTISING

Technology Levels the Field

The beauty of technology is that sometimes a **little bit of cash buys a lot of flash**. John Clarke, president of County Fair Food Stores, in Mitchell, S. Dak., explains that while his satellite broadcast system costs next to nothing, it gives his $20-million grocery business the ability to compete with the large players:

"When it comes to technology, we're right up there with the big boys. We pay only $30 a month for a satellite system. It receives broadcasts from manufacturers—commercials and discount information specifically designed for products sold in our stores—and plays them over our speakers. The broadcasts in our two stores help push items that otherwise might not move off the shelves fast enough. And the system saves us money because we don't have to pay to produce the broadcasts. I know big companies that have the same system.

"Also, because we have computers and laser printers, we can create signs and price tags in-house. The bigger companies contract to outside sources to make their signs, but ours look just as professional. We've programmed the computer so that all we have to do is enter the product information to output a sign. The system then prints out signs that meet size specifications we've previously set. It used to take us hours to make signs by hand. Now it takes a few minutes.

"Back in 1984, when County Fair Food Stores was founded, we were probably 10 years behind the big stores technology-wise. Now we're only one or two years behind. We've closed the gap because the price of technology has dropped so much."

ADVERTISING

Japanese Customers on Your Web Site

William Hunt struck out when he tried to market his earthquake kit in Japan's largest department stores. So his wife, a native of Hiroshima, helped him develop a Web site in Japanese. Now Hunt has moved into a new business. He helps **small companies market online to Japan**, in Japanese.

Hunt's company, Global Strategies, in Alhambra, Calif., helps clients avoid the mistakes entrepreneurs often make when they try to sell to the Japanese. Forget about scoring big with an English-only Web site. "You wouldn't market a car in Japan using an English ad campaign," he says. Also, he notes, people too often overlook Japanese-only Web search engines, and it's not hard to see why those listings are critical. Hunt encourages clients to state their return policies on their Web site. The Japanese are understandably leery of overseas companies, and an explicit return policy seems to ease their worries. Hunt's translators are quick to alert companies should their logos or themes be likely to strike a sour chord with Japanese sensibilities. Finally, Hunt may well suggest that his clients have mascots. Japanese audiences, it seems, respond positively to the use of symbols or characters.

Hunt cautions that if your product is not one that will fly with Japanese males in their twenties and thirties, there probably isn't much the Web can do for you. What are the hottest Web-marketed products in Japan? Outdoor sporting goods, computer software, popular music CDs, and gourmet-cooking items.

ADVERTISING

Your Brand Is Your Edge

If you rely on technology for your competitive edge, you shouldn't allow yourself to get too comfortable. Technology changes way too quickly. Ian Morrison, formerly the president of the Institute for the Future, in Menlo Park, Calif., and author of *The Second Curve: Managing the Velocity of Change* (Ballantine, 800-733-3000, 1996, $25), explains that **branding is key to keeping ahead of technology**:

"Brands will become more important as the concept of brand gets redefined. In the future it's going to be more and more difficult for a high-tech company, for example, to sustain its success. Companies aren't going to be able to sustain the technological competitive advantages they've had in the past. That's why branding the product is going to be crucial to success. Branding explains the success of a company like Intuit, which has always managed Quicken as a brand business. Successful companies grow through their ability to create value in the mind of the end user."

ADVERTISING

Establish Your Online Image

The average consumer in the average Web store has no history with the host company. More likely than not, he or she may have been surfing along and landed there purely by chance. Few Web sites feature faces, voices, or even names. The site server could be anywhere today and somewhere else tomorrow.

It's no wonder that most consumers stick with recognizable brands. Virtual Vineyards, an online wine merchant, battles the **online-credibility issue**, establishing its brand in three ways. First, the site has a strong voice tied to a specific person. Most retail sites have links to reviews by outside experts, mailing lists, newsgroups, and the like. Virtual Vineyards, in Palo Alto, Calif., runs a clean line straight to company cofounder and master sommelier Peter Granoff. An acknowledged authority in his field, Granoff writes reviews and an advice column at the site.

Second, the company uses software that expresses and focuses that voice. Peter's Tasting Chart conveys a considerable amount of information. The fast-loading pages contribute an air of authority, and the consistently uniform layout reinforces the image of Virtual Vineyards as a useful source of knowledge.

Last, the site's inventory comprises wines from a range of producers. Because Virtual Vineyards isn't tied to the products of any single winery, its customers do have a channel for expressing complaints.

ADVERTISING

The Best Places

Craig Heard, president of $16-million Gateway Outdoor Advertising, in Somerset, N.J., was among the first in his industry to use a geographic information system to identify the best places to locate specific ads. Today, all 6,000 of the firm's **billboard sites are geocoded to target ad placement**. For a retailer of children's athletic shoes, for example, Gateway's computer is programmed to select placement based on proximity to schools and playgrounds. For tobacco or alcohol advertisers wary of community opposition, Gateway's computer focuses on locations respectably distant from churches, schools, and hospitals.

New and improved geocoding helped Gateway book steadily increasing revenues over the past three years, even though many of its clients were struggling to adapt to changing times. Commissions from smoke and drink ads—once the bread and butter of the billboard industry—have dropped sharply in recent years, and Gateway worked vigorously to diversify its customer base. "The mapping system allowed us to do that by showing advertisers how well we can target their markets," Heard notes.

ADVERTISING

Multimedia Mix Maximizes Impact

By combining graphics, text, audio, and video, you can elevate your sales pitch from a whisper to a shout. Here's a brief primer on **multimedia tools** and their uses:

- Because a CD-ROM can store much more information than a floppy disk (a single CD can store up to 650 million characters), it can accommodate more compelling, dynamic presentations. Mixing art, text, audio, and video for multimedia presentations brings copy to life and captures the fancy of customers.
- A floppy disk is, however, much more affordable—generally one-tenth what it costs to produce a CD-ROM. And virtually every computer has a floppy drive, whereas many older PCs may not have CD-ROM drives. Still, you need to recognize that a floppy's capacity is limited, and compatibility issues with different systems can arise.
- You can use a floppy to create a rudimentary and inexpensive multimedia presentation for a kiosk. Well-designed kiosk presentations can do everything from collecting consumer data to placing orders. That flexibility, of course, comes at a price: A quality kiosk is priced upward of $10,000.

ADVERTISING

Cavorting with the Customers

Paul Mermelstein, president of Island Hideaways, a $1-million vacation rental agency in Ellicott City, Md., was desperate for a way to grab consumers and travel agents by the throat. The company's colorful catalogs and brochures were getting lost in a sea of travel literature advertising hundreds of "paradise-spun" properties on dozens of "sun-kissed" islands. And even when his catalogs surfaced, potential customers had little patience for the pages of photos and type, no matter how sophisticated the design.

Mermelstein realized that if he wanted his company to keep growing, he'd need to make a game of selecting the perfect vacation spot. He wanted an interactive tool that, after getting customers to specify preferences like location, budget, and room size, would **automatically sift through reams of information to make the perfect match**.

He found his answer in CD-ROM technology. On a single disk, "Villas of the Caribbean," Mermelstein crammed nearly 2,500 photographs, half an hour of music, 21 maps, enough text to fill a 200-page guidebook, and mini-spreadsheets with seasonal details on 390 villas, 23 resorts, 15 boats, and assorted cottages and condos. The disk cost next to nothing to produce, and until he started giving it away, Mermelstein granted a share of every CD sale to its developer.

ADVERTISING

Tend Your Web Site—or Take a Dive

Just because you build it doesn't mean that they will come and play. That's what Do Rags, in Hewitt, Tex., learned about the Web site it set up to promote its line of hats. The $2.4-million company is all about fun. Its products include the Viking, the Yak, and the Bonehead. But Do Rags made the **fatal mistake of relying on visitors to maintain the Web site's liveliness**, and they have not obliged.

Do Rags' site feels like a summer camp where the counselors are trying to lead group singing long after the kids have left to play cards. The site features several contests, one to design a "mental hat" and the other to take a picture of someone wearing one. Do Rags promises to name one winner for each month's contest, but for several months there haven't been enough entries to name anyone. In the site's chat area, there are few messages, and the calendar of events is completely blank.

"These concepts should have turned the site into Grand Central Station. Instead it looks like a graveyard," admits Alan Wills, Do Rags' CEO and founder. "It's totally against our culture and marketing image." What happened? Do Rags budgeted $5,000 for the site and spent it all to hire a Web-design firm, leaving zero for maintenance.

"We don't see it, so we don't think about it a lot," Wills admits. Do Rags' image-boosting tactic has become a liability. "People see it and aren't getting the most positive image of our company," he says.

MARKET RESEARCH

Web Sites That Work

Odds are that by now, you've given up pretending the Internet will have no impact on your business. *CyberPower for Business: How to Profit from the Information Superhighway* (Career Press, 800-227-3371, 1996, $14.99) does mention such technical fundamentals as modem speeds and autoresponders, but Walter H. Bock and Jeff Senné's book focuses mostly on the business concerns of the Net. In broad, clear strokes, the authors explain how companies can **make money, increase sales, cut costs, improve customer service, and recruit new employees using the Web**.

The authors appreciate online technology without glorifying it. For example, they warn, even though pictures and sound are possible and attractive, you probably shouldn't let your Web site get too elaborate until you have your strategy in place. Your goal on the Internet, they emphasize, is to provide the up-to-date information that your target market values.

MARKET RESEARCH

Biblio-Tech: Take a Look in Your Local Library

When was the last time you checked out your local library? It's probably changed considerably since the days when you relied on it for term-paper research. Today's **high-tech libraries offer Internet access**, computer-file search capability, and cyberlibrarians.

The New York City Public Library's Science, Industry, and Business Library (SIBL) has developed special services for people who run small businesses. "They simply don't have the resources in-house to do the research they need," says Betty Jean Lacy, a librarian and small-business specialist at SIBL.

Patrons must call ahead of time to reserve one of the library's 82 workstations. But it's worth the trouble. Using those computers, users can browse the Internet or search nearly 100 business and government databases, ranging from Nexis to import-export collections. The 42 computers that provide access to the library's card catalog are available on a first-come-first-served basis. Got a laptop? Bring it along: There are 500 jacks to accommodate you.

The library has been a boon for entrepreneurs like Sharon Geller Metal, whose small business, SGM Design, in New York City, makes gifts. Using one of the SIBL computers, Metal searched an online version of the *Thomas Register,* and she located manufacturers for her hard-to-make wares in record time.

MARKET RESEARCH

Beware of Bugs

Jeff Bezos, president of Amazon.com, took a simple precaution when the giant online bookseller, in Seattle, was still an unknown start-up. Bezos tested his Web site before opening it up to the public at large. Before the company posted even a single sale, Bezos launched a beta test for employees and friends.

The site was live on the Web, but Bezos didn't publicize the address or list it with Web search engines. Instead, he encouraged his employees to give the address and a fake credit-card number to their friends. He wanted them to place mock orders. The six-week test period quickly mushroomed into three months as the 300 beta shoppers reported bug after bug.

"It was a huge success," Bezos reports. "We found all the major and most of the minor bugs." He now recommends **trial runs for even the smallest Web sites**. "Most people just put up their sites and learn about the bugs over time. But," he cautions, "there is a problem with that. The first people who try it may just give up and never come back." Amazon.com has grown to be one of the Web's most publicized success stories, with record sales and a successful IPO.

MARKET RESEARCH

Fast Focus with Results

Janice Gjertsen, director of business development for Digital City, an online entertainment company with offices in New York City, wanted to conduct traditional focus groups to gauge reaction to her Web-based events guide, Total New York. Experience, however, had prepared Gjertsen for vague results. As an employee of other companies, she had tried to get objective results by using a professional to moderate focus groups, but group dynamics worked against her. She knew she needed a more **reliable source of marketing information**.

Rather than scrap the focus group idea altogether, Gjertsen decided to take it online. She hired online market research firm Cyber Dialogue, in New York City, and she got results. "People were a lot more honest online than they were in our traditional groups," she says.

Cyber Dialogue provided a moderator and drew on its own 10,000-member database to assemble the group. They held the meeting in a chat room on the Total New York Web site. Gjertsen, observing from her office computer, was equipped to interrupt the moderator with flash e-mails that didn't alert or interrupt the participants.

Such online sessions are considerably cheaper than traditional focus groups. The $3,000 Gjertsen paid was about one-third the cost, and the results were available far quicker. Rather than having to wait four weeks, Gjertsen received a full report within one day. "It's so immediate and accurate," she says. "In this world, I need that."

MARKET RESEARCH

Business "Op" Opens Online

Joe Weber, CEO of Narrateck, in Brookline, Mass., **found his dream product online**. After the engineering department shot down his proposal for a new voice digital-dictation product, Weber left his employer of 13 years, took another job, and developed and patented a speed-typing software system in his spare time. Not much happened until Weber made a fortuitous discovery:

"One Wednesday evening, I was scanning a newsgroup for transcriptionists, and my eye caught a message titled, 'Would anybody buy this system?' The writer, who it turned out was president of a major technology company, described software that sounded as if it matched the idea I'd proposed at my digital-dictation job—the one that I was told couldn't be realized. I immediately e-mailed back, asking for additional information. On Thursday, the writer and I exchanged more e-mail. On Friday, we had a phone conversation. On Sunday, I flew to Virginia to check out his system.

"The stuff really worked. On Monday, we had a preliminary agreement. I would set up a company to market the product exclusively. One week later, we signed a contract. No lawyers. A month after we signed our contract, we unveiled the software at a trade show.

"It's not easy to find the right thing. This opportunity would have blown right by me if I hadn't been networked into the right market at the right time. Fortunately, staying tapped in has gotten easier. You can do it right from your desktop. You don't want to be offline when your window of opportunity pops open."

Hey, Web-ster, What's a FAQ?

If you're about to tackle the Internet, you should know these **key terms and tools**:

Browser: A software package that allows you to access the Web.

FAQ (frequently asked questions): A list of common questions and answers for a newsgroup or Web site.

FTP (file transfer protocol): A method of transferring files among computers on the Internet. It allows you to download a file from, say, the Library of Congress to your computer.

Mailing lists (or Listservs): Discussion groups that are open only to those who "subscribe" to the main address. When one subscriber posts an e-mail message, it gets routed to the entire mailing list.

Newsgroup: Any collection of posted messages on a specific topic found in Usenet. Newsgroups are also known as online discussion groups.

Search engine: A Web site, such as Yahoo!, Lycos, or Hot Bot, that helps users locate information on the Internet.

Usenet: The large collection of newsgroups on the Internet.

World Wide Web: Often confused with the Internet, the Web is only one section of the online universe. The Web is a collection of graphics-intensive "sites" and an increasingly popular place to do business on the Internet.

MARKET RESEARCH

Use the Net to Slash T&E

Eric Anderson, CEO of Art Anderson Associates, in Bremerton, Wash., realized it was time to take the Internet seriously. His $4-million engineering-and-architecture firm designs ferry vessels and port facilities around the world. Because reliable geographic surveys are vital components of his business, Anderson used to dispatch employees across the globe on fact-finding missions. A few years ago, he tried a more economical approach.

For about $20 a month and the cost of Internet software that included e-mail, Web browser, and a few other tools, Anderson started looking for information on the Internet.

He has access to newsgroups and Web sites run by overseas tourism boards. “If I saw an intensive push to develop industry or tourism in a certain area, I would try to find out who was leading that push,” he says. Then he would contact that person about building a ferry. “The Internet let us **do more preliminary research without the travel expenses** we had before.”

ORDER-TAKING

Don't Let Customers Hang Up

Every evening at closing time, the six answering machines at Magellan's, in Santa Barbara, Calif., kicked in. And every morning the machines played back partial orders and the voices of frustrated customers who objected to leaving their credit-card numbers on an answering machine. John McManus, CEO of the travel-equipment catalog company, could only guess how many people hung up, leaving no message at all. Even though company sales had yet to reach $1 million, McManus knew he had to invest in a better system.

McManus signed a five-year agreement with AT&T. The service covers a dozen phone lines, and, in addition to fielding calls at night, it handles the overflow during working hours. That was a good deal: McManus estimated that staffing for overflow coverage alone would require two more employees, costing him at least $32,000 a year.

AT&T installed its system over a weekend, and by Monday morning, McManus knew he'd made the right investment. Previously, Magellan's machines recorded no more than 20 messages during a night—mostly catalog requests and "a few brave orders." That Monday morning, McManus counted more than 200 messages, and a significantly higher percentage of them were orders. That ratio remained high. About 35% of all calls are orders, up from 10% during the old days.

"The system began paying for itself from minute one," McManus says. "Not only did orders increase, but you have to consider the customer's lifetime value." McManus's reorder rate, about 50%, is high for the industry. A **phone system that keeps customers from hanging up keeps them coming back**.

ORDER-TAKING

To Place an Order, Please Press 1

There are plenty of options for customers who want to make purchases from Christian Book Distributors (CBD), a discount seller of religious books in Peabody, Mass. They may mail in their orders, call an operator, or dial a special number to **place orders with an automated system**.

CBD's automated system, installed in 1990, uses interactive voice-response technology. Callers use their telephone keypads to enter item numbers, the customer-identification number from their catalogs, and credit-card information. A recorded voice confirms the information they enter and generates a packing list for CBD's warehouse staff.

Cofounder Ray Hendrickson says that because each order processed by a live operator costs CBD from $1.50 to $2.00, the $40,000 system paid for itself in less than six months. About 10% of the company's phone-in customers choose the automated system.

There is a trade-off, Hendrickson admits. The orders taken by machine are, on average, 7% smaller than those taken by live operators, who often recommend additional purchases to customers.

ORDER-TAKING

Online Payments: Plastic or Digital Cash?

Few consumers have embraced the idea of using digital cash. Cdnow, an online retailer in Jenkintown, Pa., has received next to nothing in digital cash since it began offering that payment option early in 1996. Many small-business owners believe that their customers are nervous about making online purchases using an ordinary credit card. But it looks like **credit-card sales are the way to do business online**. Many company owners say their customers prefer credit cards online, and it may be many years before consumers become comfortable with digital money.

For their part, retailers say they feel more relaxed about their Web businesses than they do in their traditional bricks-and-mortar stores. Anyone who buys a product online must provide a shipping address, and that information could help authorities track down a culprit whose credit-card number turns out to be stolen. "In a physical store, the credit-card thief just walks away, and I'd never be able to get hold of him," says Chris MacAskill, president of Computer Literacy, in Sunnyvale, Calif., who has decided to skip digital cash in favor of credit cards.

For Internet commerce to thrive, buyers must become as confident as sellers. "Perception about security threats is the only problem with Internet security," says Jason Olim, CEO of Cdnow. "If consumers are buying from a company with a recognized name, there is no fear factor."

FULFILLMENT

Outsource Your Order Handling

In 1994, Air Taser, a start-up manufacturer in Scottsdale, Ariz., was trying to figure out how to process orders and handle fulfillment for its air taser guns—protection devices used to disable an assailant's nervous system temporarily. Founders Tom and Rick Smith knew that they'd need inventory-tracking technology, warehouse space, forklifts, and labor. All told, they estimated they'd have to shell out $200,000 the first year and $100,000 every year after that. "And then," says Rick, "we'd have to manage the program."

The Smiths preferred to pour their money and energy into product development and marketing. Everything pointed to one strategy: **hire someone else to handle fulfillment**. The brothers did some shopping and chose Insight, a company in Tempe, Ariz., that does direct sales and marketing of microcomputer products and services. Air Taser signed a three-year contract to pay Insight 4% to 8% of each sale in exchange for its services. Insight owns the system it developed, but the Smith brothers have access to warehouse space, machinery, and people to move their product.

Professionals handle Air Taser's order technology, and Rick says that the arrangement avoids a long list of headaches. "Do we need a bigger warehouse? How are we going to track the parts? Do we need a new inventory-control system? By outsourcing," he says, "all that is somebody else's problem. We can focus on our next nifty new product."

PRODUCT DEVELOPMENT

Checking Data More Than Twice

Whether they are test-marketing a new fruit drink, beta-testing computer software, or running clinical trials on an antibiotic, companies share a common concern. To accelerate their new product's trip to the marketplace, they need fast and reliable methods for gathering and analyzing data. The foundation is a solid information system.

But for any system to be worth its price, **data entry must be perfect**. To guard against errors, Isis Pharmaceuticals, a $26-million drug development company in Carlsbad, Calif., devised this five-step program:

1. *Manual review.* Before a clerk enters a patient's information into the computer, a data-entry operator reads the record, checking for obvious mistakes. For example, a patient's age should never be 150.
2. *Double entry.* Once the information has been keyed into the database, a second data-entry operator reenters the same record into a new file. The computer then automatically prints a list of inconsistencies, and the two operators reconcile the differences.
3. *Visual verification.* A data manager compares the entered data against the original handwritten notations.
4. *Electronic verification.* A data manager runs a series of relational edit-check programs to verify that data in selected fields of the database are logical and consistent.
5. *Final audit.* When an entire clinical trial is complete, Isis managers run a final audit, checking a representative cross section of data against the original handwritten notations.

• SALES & MARKETING •

PRODUCT DEVELOPMENT

What They Do on the Web

According to a survey of entrepreneurs, the Web is the hottest place for product development and market research. Not everyone who cruises the Web is out for a joyride or trying to sell something. An increasing number of **business users are doing online research**.

A study of Internet demographics conducted by CommerceNet (a consortium of more than 500 companies committed to electronic commerce) and Nielsen Media Research (the TV folks) reveals that of the 18 million people 16 and older in the United States and Canada who have used the Web, roughly half of them have done so for business purposes. Electronic commerce, the buying or selling of goods and services online, ranks dead last in business activities. Here's how the survey breaks down:

Activity	Percentage of Respondents
Gathering information	77%
Collaborating with others	54
Providing vendor support and communications	50
Researching competitors	46
Communicating internally	44
Providing customer service and support	38
Publishing information	33
Purchasing products or services	23
Selling products or services	13

• SALES & MARKETING •

PRODUCT DEVELOPMENT

Groupware Speeds New Product Intros

Normally, it would have taken Steve McDonnell a year and a half to bring a new product to market. But the president of the $7-million natural-foods supplier, Groveland Trading Co., in Branchburg, N.J., introduced his new, organic potpie in a record eight months. The big time saver? Lotus Notes (Lotus Development Corp., 800-343-5414). Less than a year after the company installed the groupware, McDonnell says, "It's become our general manager."

The company outsources production, and Notes, which tracks such details as product sourcing, customer interest, price points, packaging development, ingredients, and consumer feedback, keeps employees informed of where product development stands. Instead of waiting for McDonnell's approval, **employees can move their segments of the project forward** as soon as Notes brings them the information they need to proceed.

"Before, I was making decisions without complete information," McDonnell admits. "But, because I was the boss, everyone listened to me. You can move down a particular avenue too fast and end up doing things twice. That's costly, because if you're not among the first ones on the shelf, you're just not in the game."

McDonnell says the software also frees him from monitoring the mundane minutiae. Now the employees have access to all the information that he previously held. "I'm really focusing now on strategic planning, exactly what I always dreamed of doing."

PRODUCT DEVELOPMENT

Innovation Is a Process

In her book, *Wellsprings of Knowledge: Building and Sustaining the Sources of Innovation* (Harvard Business School Press, 800-338-3987, $29.95), Dorothy Leonard-Barton explains how companies can better manage their strategies, practices, and expertise to stay ahead of the competition. The Harvard Business School professor of business administration offers a tip about structuring product development:

"Often, small companies scramble just to get the first product out the door instead of stopping to think about the creative process and all the things—like incentive systems, educational systems, and recognition systems—that go into it. And it's important to do that. For example, if a manager calls for a new product development meeting, he or she usually hopes that the meeting will yield creative ideas. But the manager needs to **take a few minutes to think about process**: How are we going to organize this? How much time should we spend brainstorming? When everyone knows the procedures, the manager can spend less time fighting fires because the company has an engine in place for innovation. The more time the company spends building that engine, the less likely its competition will be able to imitate its methods for success."

PRODUCT DEVELOPMENT

Different Time Zones Extend the Day

What if you open a branch office in a different time zone? Twelve years ago, when VeriFone was still a small Honolulu-based start-up with offices scattered around the United States, Europe, and Asia, it opened a software quality test center in Laupahoehoe, Hawaii. The company took advantage of the time differences, using the "extra" hours to speed new product development.

William Pape, cofounder of VeriFone, an electronic-payment processing company, explains how things work:

"VeriFone programmers in eastern time zones put software that needs to be tested on the company network when they leave in the evening; engineers in Hawaii pull it down and run the tests while their colleagues sleep. In the morning, the East Coast programmers have their test results. It's like **a graveyard shift without the upset body clocks**."

The setup is fairly inexpensive. For an office of two to five workers, Pape uses a small private branch exchange (PBX) phone system, a LAN, and dial-out modems that use standard phone lines.

PROSPECTING

Check Your Sales Leads

You can learn about your customers even before they're your customers. Employees of ESG Internet Services, in Amsterdam, do research on clients prior to making their pitches. The small Internet content provider follows up on the leads it generates at trade shows and on the Web. But before making phone calls or mailings, the five employees begin intensive Internet information searches on each lead.

"It's surprising how much information is really out there," says training manager Cheryl Gilbert. Web sites, Internet newsgroups, proprietary online services like America Online, and pricier search services like Lexis/Nexis are tools that turn up newspaper articles, press releases, and other data about most businesses.

"We can address the leads that we decide to follow in an extremely personalized way," Gilbert says. "Companies are impressed that we took the time to research them before pitching a proposal." The **Internet works as a tracking tool**, reducing the time and money required to do the research, and it also allows ESG to focus time and mailing resources on the most promising prospects.

PROSPECTING

Unravel Sales Puzzles

Evan Segal of Dormont Manufacturing, in Export, Pa., installed **sales-force automation (SFA) software** and saw results within months. The president of the $30-million manufacturer noticed that, in certain territories, a few of the company's large customers were spending considerably less than their industry counterparts in other regions. He used SFA to run a report that itemized the products sold to each customer in the low-buying group. In a matter of hours, he had uncovered a glaring problem. Most of Dormont's large customers had been purchasing the company's newest product—a safety valve—in record numbers, but the customers in the problem group hadn't bought a single one.

When Segal questioned the reps about that poor showing, he learned that he himself was to blame. He'd been so caught up in pushing the newly patented product that he'd overlooked such seemingly mundane details as ensuring that all his reps had adequate samples to distribute to potential customers. He subsequently put together sample kits for each of the low-selling reps and also wrote up individual sales goals for them.

The SFA revelations had real impact on sales. The company sold 50% more safety valves in the year following its adoption of SFA software. And overall, he says, he expects Dormont's revenues to shoot up by 25%, thanks to the implementation of the SFA system.

PROSPECTING

Lost in Cyberspace

While it's true that e-mail is a cheap way to reach thousands of people quickly, it's not always reliable. At least that's been the experience of Neodata, a subscription-fulfillment company in Louisville, Colo. The company tried to e-mail messages to 85,000 names. At last count, however, only 17,000 messages had been delivered.

The problem, according to Mark Bosché, who at one time was Neodata's director of Internet services, is that **e-mail addresses change at an astounding rate**, and there is no clearinghouse to track the constant churn. "There's a huge turnover of e-mail addresses right now, and there is no great way of keeping up with it," Bosché says.

Postal mailing addresses, on the other hand, are fairly stable, and the U.S. Postal Service sponsors a national change-of-address program that works. Of the 1.9 million pieces of snail mail that leave Neodata's facility every day, less than 1% go undelivered.

PROSPECTING

Phone Numbers—Faster

Thomas Norton remembers the bad old days when his company, Fidelifacts, in New York City, used to spend at least $2,500 each year for phone directories from all over the country, and yards and yards of shelf space were sacrificed to store them. But Fidelifacts needed the directories. The 25-employee company specializes in background investigations of prospective employees, a task that requires checking many references.

Now, the company tracks down elusive references using Select Phone (Pro CD, 800-99-CDROM), one of a number of inexpensive CD-ROMs that provide national phone and address listings. Several Web sites also offer search engines that turn up phone numbers, street addresses, and e-mail addresses. Even though some of those listings are out of date, Norton says the **electronic phone listings save time** for Fidelifacts' employees.

Many electronic phone directories let users compile customized mailing lists of likely prospects, selected by standard industrial classification code or zip code.

PROSPECTING

Automate Your Remote Sales Force

Although sales-force automation (SFA) software is a powerful tool for tracking prospects and sales, it has other simple applications.

Qiagen, in Valencia, Calif., is a growing biotech company with a 60-person sales force spread across the United States and Canada. Headquarters once used the telephone and overnight delivery services to get information about leads to its sales reps. But the whole process was loosely structured, with no formal mechanism for regular updates. All too often, the sales reps were left in the dark. They couldn't keep up with the new leads in their own sales areas.

At a cost of $1,500 for each user, Qiagen invested in SFA software and gave each of its reps a laptop computer. That seems to have rectified the problem. Today, reps **use their computers and modems to dial in for weekly reports on the latest prospects**. Hot leads no longer cool off for lack of attention, and Qiagen no longer pays a fortune to the phone company and Federal Express.

PROSPECTING

Be Ready to Strike

Sales-force automation (SFA) still means different things to different people, but the SFA software out there is powerful enough to do most of those things simultaneously. Still, rather than tackling the entire beast at once, many small companies get better results by automating one step at a time.

"Our sales cycle involves long-term commitment," says Scott Specht, president of Telepress, in Issaquah, Wash. For Telepress, a $5-million printing company, wooing a customer can take years. Two- or three-year contracts bind most Telepress sales prospects to other vendors. For quite some time, the company had assigned one employee to maintain a manila-folder filing system of correspondence with prospects and a calendar crammed with penciled-in reminders for appointments. When Telepress decided to expand its markets beyond Washington, it was plain that the antiquated system could never work.

Now, using **software designed to track correspondence**, Telepress handily follows more than 300 prospects. Because of the scheduling function that automatically sets dates for follow-up letters and phone calls, Telepress sales reps are ready to strike just as prospects' old contracts come up for renewal.

PROSPECTING

Are You *Really* Ready for the Web?

It's one thing to get all excited about the Web," says Karen Rizzo, marketing manager at Kryptonite, a manufacturer of bicycle locks in Canton, Mass. "But **whether or not it makes sound business sense to be on the Web** is a completely different matter."

Rizzo knows that typical Internet users are upscale, well-educated young men, like her bike-lock customers, but she has postponed selling on the Web until she can pinpoint her costs and calculate the potential return. After all, a simple Web site can easily cost thousands of dollars to develop. Equipping a company for online sales transactions and customer interaction can cost considerably more.

Kryptonite, which sells through distributors, tested the waters with a small site that offers press clips about the company. Visitors who request product or company information get it through the regular mail. "We can track how many people ask for information and whether their requests generate business," says Rizzo, who budgeted $10,000 for the site's first year. Her next step will be to interact with customers online by hosting e-mail focus groups about emerging products.

Rizzo expects that the extra research time will pay off. "Lots of companies establish a Web site and then pull the plug, because they didn't anticipate how to do it well, and the costs and time demands for setting it up and maintaining it got out of hand," she says.

PROSPECTING

Automate Salespeople to Keep Them in the Office

For many companies, sales automation means installing software on the laptops of their traveling salespeople. But at Lori Sweningson's JobBOSS Software, technology brought the sales force home. For most of its history, the Minneapolis company sold its shop-floor management software through informational seminars and visits to small metalworking companies. Now, with the help of technology, the $8-million company has put sales back in the home office, and Sweningson says productivity is up.

JobBOSS uses the telephone to book sales, which, on average, are worth $14,000. While a salesperson in Minneapolis and a job-shop owner anyplace in the country converse on the phone, the salesperson types information about the potential customer's organizational structure into a computer. After the phone call ends, the salesperson faxes the description to the prospect. That fax immediately demonstrates that the salesperson understands the customer's needs and how JobBOSS can help.

With the help of Close-up (Norton-Lambert, 805-964-6767), a software package that allows remote control of a far-off PC through a modem connection, JobBOSS conducts **long-distance product demonstrations** as well. The salesperson demonstrates JobBOSS software on the prospect's screen while describing steps over the phone.

By reducing travel, the new system has increased productivity 45% while reducing costs. Sweningson says that the average sales cycle has shortened from 5.2 to 3.1 months.

PROSPECTING

Build a Customer Database That Works

Here are five tips to consider if you're thinking about **automating your customer database**:

1. *Pinpoint your objectives.* Do you want to expand your customer base? Do you expect to double sales of a product line? Do you aim to encourage repeat business? Make sure you understand your goals before you automate. "Companies that are drawn to the technology without knowing what they want to get out of it will be very disappointed," says David Shepard, of the consulting firm David Shepard Associates, in Dix Hills, N.Y.
2. *Choose the software that best fits your goals.* If, for instance, you need a system that is tied tightly to sales-rep activity in the field, look for a contact-management program with database capability. If, however, you do a lot of selling over the phone, you may need a telemarketing program.
3. *Determine what to include in your database.* Once you decide which kinds of customers have the most potential, you should construct the database, mining everything from salespeople's contacts to lists you purchase from outside suppliers.
4. *Develop a plan.* Only after you define your objectives and construct your database is it time to devise a marketing program.
5. *Measure results.* Generate periodic status reports that detail such information as cost per contact and cost per sale. To be really careful, don't launch your program at full tilt. Start with a small prototype, and if you like what you see, expand it to include the entire database.

PROSPECTING

Kiosks with Sizzle

To help customers take advantage of its fancy cooking implements and exotic ingredients, gourmet retailer Jensen's Finest Foods, in Palm Springs, Calif., installed a kiosk in one of the chain's six stores. Floor traffic hadn't been a problem, but management wanted to **encourage customers to purchase more** of the store's offerings. Jensen's leased a text- and graphics-based kiosk, which was such a success that eventually Jensen's bought it outright. For a small fee, CompuCook, its kiosk developer in San Francisco, supplies software updates.

The kiosk's attraction for Jensen's affluent and sophisticated clientele is its database of recipes developed by many of America's leading chefs. The customers also appreciate being able to print copies of the recipes to take with them. Every month features new recipes, which, because they are based on input from Jensen's, highlight ingredients the store wants to promote.

PROSPECTING

Groupware Gives Quick Access to Client Data

One picture on a computer can be worth a thousand words. That's what Greg Voisen, owner of North County Financial Associates, a financial-planning firm in Vista, Calif., discovered. He hooked up his eight employees with a groupware program that presents information visually, showing relationships among clients, prospects, projects, and salespeople. Over the course of six years, the $1.5-million company had tried various programs to track the status of salespeople's work with prospects, but Outlook (Microsoft, 800-426-9400), in combination with the company's own software, made the difference.

Say Voisen is trying to sell a package of financial products to a local manufacturer. All the information on his 6,700 prospects—names, addresses, past conversations, letters, memos, and contracts—is filed away and cross-referenced under various headings. But what his employees appreciate is being able to click on a file and bring up a **visual display of links among client contacts**: salespeople, the prospect's employees, other clients, past projects, and related documents. If the display showed that the prospect had worked with one of Voisen's long-term clients, a salesperson might call that customer and ask him to put in a good word for the company. Or if North County Financial had made a proposal to the client a few years earlier, Voisen could pull that document and read it.

Voisen says that the software has dramatically improved his ability to keep track of his top 40 clients and top 40 prospects. Since he first started using groupware, Voisen has seen his business average 7% annual growth.

PROSPECTING

Target Customers Where They Live

The new generation of geographic information systems (GIS) software has worked wonders for Archadeck, in Richmond, Va. The fast-growing, $25-million home-add-on franchiser has nearly 70 offices in 21 states. The company first used **GIS prospect-finder software to manage direct-contact marketing campaigns**. Before that, Archadeck tried drumming up business the old-fashioned way: The sales force hung brochures on doors, put placards at construction sites, even made cold calls at houses that looked like they could use work.

Now, every week, the company logs as many as 120 new projects into the GIS database. Drawing a circle around each one, the company's marketing director instructs the computer to retrieve names, addresses, and telephone numbers of the surrounding homeowners.

The software extracts a list of prospects. The next step is to apply that information to a direct-contact campaign. The company uses a method employees know as "geoneighboring." Before construction on a new project gets underway, Archadeck mails a soft-sell notice to a number of "high-profile, highly probable prospects" in the immediate vicinity. The notice encourages the neighbors to phone if the construction makes too much noise. A second postcard sent shortly after construction begins is more suggestive: "We're adding a new deck to 100 Chestnut Street, a couple of doors down. Why not come over and have a look?"

PROSPECTING

Irresistible Telemarketing

Your telemarketing computer can save you time and money when it initiates sales calls. But why program it to say something boring, like, "Please hold for an important message," or "We have some exciting news for you!"?

Instead, make important customer information part of your sales lead. Here's an idea from a metropolitan police department. In Oxnard, Calif., David Keith, community affairs manager at the local police department, discovered a pattern of break-ins in a local neighborhood. "The guy was going in through sliding glass doors at the same time of day, and he was stealing only VCRs," Keith says. The department had just paid $12,000 for a complete telemarketing computer system: a PC with voice-processing boards, an uninterruptable power supply, a printer, a phone, and telemarketing software. **The system can be programmed to deliver a prerecorded message to phone numbers** stored in the database.

Keith programmed the telemarketing computer to call every home in the neighborhood, saying "This is the Oxnard Police Department with a crime-alert bulletin for your neighborhood." Keith says, "I don't think people hang up when they hear that." The computerized message described the crime pattern and recommended that residents lock all sliding and garage doors. The message also included phone numbers to call with information about burglaries or to find out how to start a neighborhood crime-prevention program.

Log On for Government Contracts

When government agencies and military bases need to procure goods, they submit requests in writing to the U.S. General Services Administration (GSA), which posts the bids in such publications as *Commerce Business Daily*. Small companies, like CRC Products, a 62-year-old, $7-million food-service equipment distributor in Terre Haute, Ind., look to that paper for opportunities. The company also solicits the GSA directly, continually updating the agency on CRC's offerings. In the past, CRC would draw up a bid and either fax or mail it to the GSA. If the agency accepted the bid, it would return a purchase order. The process could take as long as a week.

However, using a modem-equipped computer and a private network connection, CRC now has **access to government "electronic bid boards."** Although the GSA still requires signatures, many government agencies post requests for goods electronically. Every day, CRC downloads the government's requests from its electronic mailbox.

To expedite the process, CRC filed an electronic registration form with the government, marking off the categories of supplies the company offers. Each category corresponds to a different stock number. The stock number is used to sort incoming government bids and filter the pertinent ones to CRC's mailbox. Not only is CRC privy to more bids, but because the company can secure contracts faster, it has more time to fill orders.

• SALES & MARKETING •

PROSPECTING

Let Leads Qualify Themselves

Most trade-show handouts end up lining hotel trash cans. How can you ensure that your pitch isn't getting tossed or lost in the crowd? Sales reps from Imedia, a marketing company in Morristown, N.J., set up **a fax-back service at each trade show**.

They take two laptop computers to each show—one loaded with database software, the other with fax software and a modem. Prospects type their own names and addresses into the company's database, then move to the second computer, where, from an on-screen form, they select those items that interest them. Imedia faxes the material directly to the prospects' offices back home. It pays about $300 for trade-show phone lines and less than $1 per fax.

People who request the information that way are usually very good leads, because they've invested their time on the computers rather than merely dropping business cards into a bowl. Meanwhile, the company builds its database more quickly than if reps had to depend on someone back at the office to enter the information into a database.

Bad PR, Cyberstyle

One thing is certain. When you establish an online presence, your **competition will be watching—and perhaps contributing to—customer complaints**. Without question, most of its competitors were watching when Polaris, in Escondido, Calif., released a buggy version of PackRat, its personal information manager software. Competitors were monitoring the company's online forum on CompuServe before, during, and after its troubled period. CEO Jack Leach suspects that at least one of his competitors publicly attacked Polaris online.

One of the most distressing aspects of the ordeal was the presence in the Polaris online forum of numerous conversations that not only derided PackRat, but even extolled the virtues of competing products. If employees of a rival company promote its product in your online forum, that's considered "poaching." If ordinary citizens promote a competing product, it isn't poaching. But, Leach points out, given the ease of assuming identities online, poachers can pretend to be innocent observers and do equivalent damage.

And, of course, the press will be watching. The press is probably what did Polaris the most harm, according to David Coursey, a software-industry observer. "The real damage was done by the people who never saw the messages themselves but became convinced that Polaris was in trouble." Customers faulted Leach for not confronting the online situation much sooner than he did. "Leach took his sweet time responding," says one angry former customer. "I'll never buy a Polaris product again." Being online didn't cause Polaris's problems, but it did quickly exacerbate them.

PUBLIC RELATIONS

The Medium Is the Message

The most dramatic changes to the public relations game are in communications. E-mail and the Web are important new tools, but they can easily be misused. You need to **apply new media to media relations**. First, remember to reserve a spot for reporters on your Web site. You should include basic company data, a company history, and the most recent press releases. Feel free to direct media contacts to the site, and expect that anyone researching your company will look for your site. Second, remember that not every reporter relies heavily on e-mail. Many prefer to receive unsolicited information via fax, telephone, or mail. Let your press contacts tell you the method they prefer.

And don't forget to check out new media wire services. For a newspaper publicity photo shot, Pete Slosberg, founder of Pete's Brewing, in Palo Alto, Calif., posed in a bathtub surrounded by his liquid assets. But rather than simply mailing the photo and press release, Slosberg's publicity pro, Kristin Seuell, filed the photo with PhotoWire, Business Wire's commercial service that sends photos digitally into the darkrooms of more than 500 newspapers as well as ABC and CNN.

The color photo, with its caption, "Specialty brewers make it from bathtub to big time," was picked up by 35 papers across the country, and Seuell says the $725 filing fee "was worth every cent."

PUBLIC RELATIONS

Satellite Dish for a Day

Drew Santin was excited to learn that his trade association was cosponsoring a seminar that would be transmitted nationwide via satellite. The conference topic—using technology to make prototypes rapidly—described his business exactly. However, there was no place to view the telecast near Santin Engineering, in West Peabody, Mass.

For a $400 fee, paid to the conference producers, Santin transformed his $4-million **company into an authorized conference-viewing site**. He paid $700 to a nearby satellite-dish rental firm for one day's use of a dish, including installation and setup. Santin could have held the event in a hotel, but because he welcomed the chance to show off his plant, he invited customers to watch the seminar and enjoy a buffet lunch and a company tour. The company also sent out press releases about the telecast.

Although guests were given short notice, Santin estimates that more than 60 people attended, about half of them current customers and half prospects. During the show, Santin collected questions from the audience and faxed them to the speakers. He later addressed any questions that the panelists hadn't.

Santin is pleased that his company got to showcase its expertise. In addition, the press releases raised the profile of the 42-year-old family business and resulted in several interviews by local radio and cable stations. Santin offers only one caution to others considering a similar event: Make sure the satellite link is properly tuned. When the broadcast began, Santin's guests found themselves watching a fishing program.

PUBLIC RELATIONS

Douse Flamers Online

You can **go online to combat negative publicity** about your company or products:

- *Act immediately.* You may be able to keep a few negative comments from igniting a flamefest.
- *Involve top-level management.* When the situation is potentially critical, you need to develop a high-level strategy and ensure that all company representatives are speaking with the same voice.
- *Promptly assume responsibility for any errors you've made.* Do everything you can to rectify those errors, and keep customers informed. If the problem is one of misinformation, keep posting the truth in as many places as possible until it sinks in.
- *Act aggressively to involve the press.* Spread your version wherever possible.
- *Hold your temper.* No matter how provoking a flamer is, your company's response should always be courteous and professional.
- *Recognize that flaming is a symptom, not a cause.* Companies usually don't get flamed without reason. Taking good care of your customers is the best way to prevent flaming.

III

• MANAGING PEOPLE •

"The biggest impact
of technology has been
to allow us to do unproductive
things at a far more
impressive rate…I can write a
useless memo and e-mail
it to dozens of people
who are too busy writing
their own memos
to do anything about mine."

SCOTT ADAMS
creator of Dilbert, Dublin, Calif.

COMMUNICATIONS

Weekly E-mail Updates

As companies get bigger, and employees move in different directions, it can be difficult to keep track of how their projects are progressing. One way to stay in touch while keeping the organizational structure flat is to have staff **write short e-mail updates of each week's work**.

At Indus International, a management-software company based in San Francisco, all 930 employees submit one-page summaries each Friday. Their reports outline their accomplishments of the previous week, their plans for the coming week, and impediments they have encountered. Having to formulate such messages helps employees organize their work and pinpoint problems. The weekly packet is posted on the company's electronic bulletin board for everyone to read.

One person who reads the entire collection is CEO Bob Felton. He also responds to many of them. "It helps short-circuit what happens in hierarchies where you're not allowed to go and talk to someone," says Felton, who did stints in the Navy and as a *Fortune* 500 executive before starting Indus. "And because people have to do it, it's not like they have to take a risk to go around a chain of command. It's what we expect."

COMMUNICATIONS

High-Tech Suggestion Box

If your employees work at networked computers, you can take advantage of the link to create an updated version of the old suggestion box. The new version: an **online suggestion database**.

At Timeslips, a software publisher in Dallas, chief operating officer Mitchell Russo has established a suggestion database, and anyone can submit a suggestion or peruse the postings of other contributors. Russo reports that the database "probably has more than a thousand suggestions in it." The ideas come from customers, letters, developers, and staff. Ranging from fleeting thoughts to formal ideas about product enhancements, each entry includes the idea, who originated it, and the date of the suggestion.

Russo reads through the collection every week or so. "About 15 or 20 of the items are what I consider great ideas," he says. "They end up on a short list in my personal database." They also show up in product updates. Russo says that more than 100 electronic suggestion-box entries were incorporated into one product release, and he has overseen development of new products that bubbled up through the idea-exchange system.

COMMUNICATIONS

All Talk and No Action?

Is technology helping us work faster and better, or is it just wasting our time? The telephone does both. In a 1997 survey, the American Management Association found that most small-company CEOs spend one to three hours a day on the phone.

Jim Iversen tries to make the best of things. The general manager of W&H Systems, a $38-million integrator of material-handling systems in Carlstadt, N.J., says he often spends several hours a day on the phone. "To get anything done," Iversen says, he usually gets to work at about 6:30 a.m. and frequently stays past 5:00 p.m. Hoping to **avoid the barrage of calls**, he works at home every Monday, but he does call in every hour to check his voice mail.

Phone Habits of CEOs	Percent of Respondents
How much time do you spend on the phone daily?	
Less than one hour	18%
One to three hours	68
Three to five hours	11
More than five hours	3
Do you answer your own phone?	
Yes	41%
No	52
No answer	7

Phone Habits of CEOs	Percent of Respondents
How quickly do you return calls?	
Same day	43%
Within 24 hours	30
Within 48 hours	5
Other	22
Do you place your own calls?	
Yes	97%
No	3

Source: American Management Association

COMMUNICATIONS

No Business Too Small

Intranets are sprouting up like weeds among the *Fortune* 1,000. Employees praise their democratic character, and executives extol their incomparable return on investment. But what role, if any, will they play in smaller companies? Steve Telleen, coiner of the term "intranet," is founder of Iorg.com, an intranet consulting company in Pleasanton, Calif. He suggests even small organizations can benefit from having intranets:

"Intranets—and this is true for companies of any size—can help people communicate when they have trouble getting together in one place at one time. If you've got people who are not all working on the same schedule and on the same projects—who have things pulling their time apart—an intranet lets them collaborate asynchronously.

"**The most successful intranets start at the grass roots**, where you don't have a lot of rules and policies about publishing. It's best to establish a seed period during which you encourage as many nontechnology people as possible to publish. Sure, you can put up a company phone directory if you're big enough, or marketing material, or meeting minutes. But I say, let people start using the intranet as a way to present to others what they're doing in the business. And it will evolve out of that."

COMMUNICATIONS

Intranet Pros and Cons

Unlike many hyped-up trends, **intranets make sense for small businesses**. These internal Web sites provide an easy-to-reach and easy-to-update way to store the information employees require. Here are some reasons to create an intranet:

- Your business uses several different types of computers—and the users all need to reach the same company information.
- You need to centralize data in an easy-to-access way.
- You'd like to give employees controlled use of more of the information stored on your network, but…
- You don't want to administer network versions of various software packages.

And here are reasons *not* to create an intranet:

- You don't have a local area network, or your network's operating system—like some older versions of Novell NetWare—does not use TCP/IP, the Internet communications standard.
- You don't have any compelling business justification for one.
- Many of your employees don't use computers.
- Nobody is available to set up and manage the intranet.

COMMUNICATIONS

E-mail That Won't Be Ignored

Do you have an urgent e-mail message for all your employees? Check to see if your e-mail software can be programmed to demand a confirmation once people have read an urgent message. At Emerald Dunes golf course in West Palm Beach, Fla., president Ray Finch has software that does just that.

Finch's software tracks tee times and players' progression through the course. Each cart has a small monitor mounted on its rooftop for players to watch. A color-coded map blinks white when a cart is ahead of its target time, yellow if it's behind, and red for very slow play. Should a foursome be holding up the players behind them, the software lets the pro shop send a canned message: "Please maintain pace with the timer." A staffer in the shop can also type in a special message. Golfers receiving a message either press a button to **acknowledge the e-mail or stare at a blank, disabled screen**. For most players the electronic nudge is preferable to getting flagged down by the dreaded course ranger.

COMMUNICATIONS

You Can Reach Me At...

Don Rose, cofounder of Rykodisc record company, in Salem, Mass., relies on e-mail to stay in touch with employees. While he does use e-mail to stay in touch with business colleagues and artists outside the company, his **highest priority is to be accessible to insiders**. With some employees, he says, he has almost no face-to-face interaction, but they have become regular e-mail correspondents. "Some people are just more comfortable leaving me an e-mail message than walking into my office."

Others in the company rely more on the telephone. "There is a certain anonymity with e-mail, and there seems to be less personal responsibility for what you write than what you say," one manager explains. Managers who find that even voice mail can be too impersonal go to some trouble to bridge the high-tech/low-tech gap. They craft a personable outgoing message, encouraging employees to stay in touch.

COMMUNICATIONS

Videoconferencing: Seeing and Believing

Thinking of adding videoconferencing capabilities to your business computers? It can save you money if your company regularly rounds up far-flung employees or clients for meetings. And it allows you to check in with telecommuting employees. But before you rush to invest in phone lines and video-capable computers, make sure to **test-drive a few videoconferencing systems first**. And to help decide whether videoconferencing is right for you, bear these guidelines in mind:

- If face-to-face contact isn't an essential part of your business transactions, videoconferencing is probably a waste of your money.
- Be realistic. The video images you'll get from a low-end desktop system move at about half the speed of television.
- Most systems don't yet talk to one another. Make sure the party with whom you aim to do business is equipped to take your call.
- No matter how well it works for you, never let videoconferencing replace all of your one-to-one meetings.

COMMUNICATIONS

Should You Buy Groupware?

Is groupware for your company? The organizational software can be a lifesaver, tracking almost all of your company's information for easy employee access. Or it can be way too much, smothering everyone with an overabundance of confusing data. To help decide **whether or not to invest in groupware**, study these possible applications:

- *Knowledge sharing.* You can store information in just about any form and easily search in one accessible place.
- *Group calendaring and scheduling.* Anyone can check schedules and set up meetings.
- *Real-time meetings.* Participants can be linked together in a network over which they can answer questions, make comments, and vote—all anonymously.
- *Bulletin boards.* Participants can carry on "conversations" over long periods of "non-real time." All comments are stored and organized in easy-to-retrieve form.
- *Group document handling.* Sitting at their own computers, several colleagues can work simultaneously on the same document.
- *Work flow.* Analyze processes by tracking the status of documents—who has them, who has fallen behind, and who gets them next.

COMMUNICATIONS

When E-mail Isn't Enough

Attorney Bill Wright knew he had problems. In fact, what he had was a failure to communicate. First, one of his three partners up and left, taking with him 17 of the Bellmawr, N.J., law firm's employees and lots of clients. Then, within days of his departure, the runaway lawyer sued his former partners, and Wright's $2.6-million firm turned around and filed a countersuit. What with impromptu hallway discussions, emergency meetings, news flashes, urgent requests for background, and the rest of their caseload, the remaining staff at Farr, Burke, Gambacorta & Wright barely had time to breathe. "We had to find a way to help us **handle the flood of internal information**," says Wright. "And we had to find it fast." Simple e-mail, he knew, would not suffice. He needed something that would organize and catalog information, not just zap bulletins around the office.

Wright and his colleagues turned to groupware, a category of software that helps numbers of people work together effectively. Instead of holding one meeting after another, the firm bought and installed TeamTalk software (Trax Softworks, 800-367-8729) on its network and started using it to discuss business. When Wright had to produce a draft of a proposed settlement, for example, he no longer had to rummage through notes from previous discussions or interrupt his partners with time-consuming questions. He simply checked the program's database for a record of earlier memos and messages relating to the topic.

A year after the suits were settled, groupware was well established as the interactive glue holding Wright's firm together.

RECRUITING

Candidates Click with Online Talent Search

If you need a new way to recruit workers, why not search online? Mark Zweig is the founder of Zweig White & Associates, a $3.1-million publishing and consulting company in Natick, Mass. Like many other employers, he posts job openings on his company's Web site. But that's not all. Instead of waiting for applicants, Zweig takes a more aggressive approach. He logs on to America Online (AOL) and takes advantage of the information available in AOL's member directory.

People who subscribe to AOL have the option of submitting a biography that includes their hometown, job description, the kinds of computers they use, and even a personal slogan or quote.

When Zweig does an online search, he clicks in to the directory and creates a list of AOL subscribers from cities and towns in his company's vicinity. He searches those members' occupations by keyword and **sends e-mail to people with job descriptions he finds appealing**. "It's much faster than a Sunday newspaper ad," Zweig explains. And with AOL's tools, Zweig can focus his search more easily than he could if he had to paw through a pile of résumés.

Zweig says that when he sends people e-mail asking whether they'd like to work for him, they are pleasantly surprised.

RECRUITING

Recruiting from Near and Far

David Minor, president of Minor's Landscape Services, in Fort Worth, took an aggressive **direct-mail approach when he recruited for two technical positions**. "It's difficult to recruit locally, especially when you need someone with specialized technical skills or someone licensed," Minor says. "It's not like there's one on every street corner."

He located a publication that lists Texans who have the required licenses, and he sent a letter explaining the job to everyone in that book. The landscape-maintenance company, with annual sales exceeding $6.5 million, quickly identified more than 20 qualified candidates and filled the positions.

Richard McCarty, a principal of the 25-person McCarty Architects, in Tupelo, Miss., has faced similar problems finding qualified architects in rural Mississippi. He gets good results with a lively direct-mail recruiting package he sends to 40 chapters of a national architects' association. "It's hard to get people fired up about Tupelo," McCarty says. "We try to throw in something that will catch someone's eye and give an indication of the spirit of our firm."

RECRUITING

Recording for Your Protection

Concerned about your interviewing techniques? Many personnel consultants recommend **audiotaping job interviews**. They say that candidates answer more thoughtfully and honestly when they know that you're recording every word. Of course, the law requires you to ask interviewees for permission to tape.

The technique worked for Drew Conway, CEO of the Registry, a provider of information-technology consulting in Newton, Mass. Conway says that taping the interviews not only helped assess candidates, but also helped him hone his own interviewing techniques. He began to notice how much talking instead of listening he did, and he learned how to pose open-ended questions that elicited more than yes-or-no responses. Asking candidates to provide examples that supported what they were saying also strengthened the results of each interview.

To stay out of legal trouble by taping, it's imperative to treat all applicants the same. You want to avoid being accused of discrimination, so if you tape one interview, you must tape them all. And hang onto those tapes. They may protect you from unfounded claims of discrimination by providing proof of equal treatment in hiring.

RECRUITING

How to Find the Right Consultant

Don't forget that job applicants aren't the only potential hires you need to evaluate. Consultants count too. Here are six tips **for interviewing and hiring computer consultants**:

1. Widen your search. Good consultants will travel, and many can work on your systems by modem.
2. The best referral system is word-of-mouth. Ask local executives and trade associations for recommendations.
3. Don't assume that consultants are good just because they've been around for a while. A bad consultant can make a living messing up one job after another.
4. Check references carefully. Don't just call. Visit former customers of potential hires and inspect their work. If a consultant always recommends the same solution, you're probably not dealing with an independent consultant, but someone who profits by selling specific products.
5. During interviews, pretend you're hiring an employee. People skills and communication skills matter—you don't want a lingo-spouting chiphead who'll alienate everyone in sight.
6. Ask about professional affiliations. Membership in certain groups means the applicant subscribes to a code of ethics.

HIRING

Site After Death

When a company goes out of business, or even when it's only downsizing, the rank and file say their good-byes and begin job hunting. Technology can help released workers find new jobs. Before it declared bankruptcy, Nets Inc., in Cambridge, Mass., was a Web-based company specializing in online commerce. Two days after the CEO, Jim Manzi, made the announcement, software engineer Kevin Jarnot **created a Web site to help the 200 or so laid-off workers find new jobs**. "When we got Manzi's good-bye letter, we broke out the Scotch and started commiserating with each other," recalls Jarnot. "After a few glasses, I volunteered to build the Web site."

Jarnot created a bulletin board where recruiters could post job opportunities and employees could ask questions like, "Will I receive my last paycheck?" He also compiled a directory of employees' names and gave his former colleagues the option of posting their résumés online.

The Web site quickly became a hot spot for headhunters. "I would put down the phone, and there would be three more calls from recruiters," says Jarnot. Today Jarnot, along with most of Nets' former employees, has returned to work in the Cambridge office of the company's new owner, Dallas-based Perot Systems. But the site is still up and running.

Step Two: A Faxable Application

Finding the right international distributors is among the most difficult—and crucial—steps a company takes as it moves into the global marketplace. It's critical to assess how involved and productive a potential rep will be. Here is one surprisingly simple move that can help weed out at least half the bad candidates: **Send potential reps a one-page fax-back application**.

Electronic Liquid Fillers (ELF), in La Porte, Ind., started using such a form in 1990. Foreign reps learn about the company at trade shows or from advertisements in trade magazines. They usually take the first step, phoning or faxing requests to represent the company. ELF then overnights packets of company information, including equipment spec sheets and the application form. The form, according to ELF's former vice-president of sales and marketing, Jeffrey Ake, "dictates seriousness."

ELF receives about 10 inquiries a month from reps, many of whom have job assignments already in mind, and from those, the company screens out people working for direct competitors, looking for small companies that will give ELF focused attention. Registered reps are granted one-year account exclusivity for their prospective clients and leads.

In 1990, international business made up less than 15% of ELF's sales; in 1997, it accounted for 25% of the company's revenues.

HIRING

Call in the Professionals

Brendan Moylan was dissatisfied. Moylan is chief operating officer of a $38-million direct marketer of soccer and lacrosse supplies, Sports Endeavors, in Hillsborough, N.C. The fast-growing *Inc.* 500 company uses mail-order software to track orders, customers, and inventory. For years, Moylan had relied heavily on his software vendor's expertise and technical support. But when he became dissatisfied with the service he was getting, he decided it was time to switch.

The ensuing software conversion, says Moylan, "was an absolute nightmare." Like many other young companies, Sports Endeavors had been doing without a director of management information systems (MIS). During the conversion, Moylan found out the hard way that **he needed in-house technical expertise**. Although he had scheduled the change for the slow season, the transition dragged on and overlapped with the World Cup soccer playoffs—the company's busiest time ever. When the new system went live, it directed employees looking for products to the wrong areas of the warehouse. The company fell two weeks behind in order fulfillment. Employees were stressed-out, customers were angry, and Sports Endeavors lost some $3 million in business.

Now, with Moylan vowing to develop more in-house expertise, the 325-employee company has an MIS director and an MIS department. As for the former software supplier, it appears Moylan made a prescient, if painful, decision. About a month after the conversion decision, Sports Endeavors got a fax from the old vendor—announcing its abrupt withdrawal from the software business.

HIRING

Telecommuting Compensates for Labor Shortage

Donald Fay, operations manager for the Payne Firm Inc., a $3-million environmental consulting firm in Cincinnati, faces a labor shortage. In the Cincinnati region, where the unemployment rate hovers around 4.5%, a full point below the nation's average in 1997, finding qualified workers has been a real challenge. So Fay has turned to **technology that taps an underused labor source**: stay-at-home moms.

To attract his new labor pool, Fay, somewhat reluctantly at first, introduced telecommuting at his company. "Six years ago," says Fay, "it would have been hard for me to get past the idea of paying someone who's not in the office." But after spending thousands of dollars in recruiting-firm fees, he realized that he could get better results by offering better options at his company. The flexibility of telecommuting enabled him not only to hire new workers, but to keep valuable employees, including his own secretary, from leaving. Fay intends to provide the option to more employees.

HIRING

Wanted: Model CIO

IQI, a database marketer in Los Angeles, was growing fast, and there was no time to make mistakes. "We required a very complex multimillion-dollar telemarketing system and couldn't afford to hire the wrong person to handle it," says founder and chairman Robin Richards. The headhunters he contacted lacked the expertise to help him, and when Richards read applicants' résumés, the technical jargon made his eyes glaze over.

"I'm a marketing and finance guy; I don't know technology," he concedes. When he saw an issue of *CIO* magazine, he realized that he needed someone like the people featured on its cover. It dawned on him that **magazines would be a great source for potential hires**. He could sidestep headhunters and stop wasting hours checking out unqualified applicants.

After an hour scanning the covers of the 45 computer magazines he receives, Richards found articles about people who had built systems similar to his. Over the next month, Richards contacted six story subjects before he offered the position to the man who accepted and oversaw the integration of the new system—without a hitch.

Richards enthusiastically endorses his headhunting technique, which he has used successfully to fill other positions. "It's probably the greatest recruiting tool for a principal," he says. "It's also about the cheapest."

HIRING

Student Interns: High-Tech, Low-Cost Consultants

President Dave Myhr's customers were asking him for e-mail and Internet access, but Specialty Engineering, in St. Paul, Minn., was not wired. The $7-million sheet-metal fabricator didn't have the resources even to research different Internet options. Paying his engineers to surf online would have been a huge salary drain, and hiring a consultant would have cost at least $75 an hour, not including Internet connection costs. He solved his problem by hiring an intern.

Myhr interviewed and **hired a $10-an-hour student intern** who had learned about the Internet by using it in college. The intern researched Internet accounts and determined which would be the best for Specialty. Myhr described his goals to the student and regularly checked on the progress of the intern's research. Myhr was prepared to take the project in-house if the intern didn't progress as fast as Myhr wanted.

Within three weeks, however, the intern had recommended an Internet account with a local service provider. The total cost: $1,280 for the intern and $25 a month for unlimited full Internet service, including an in-house connection, a company e-mail account, and Netscape's Web browser.

"Even if I'd hired a consultant and the learning curve were cut in half, it still would have cost me two to three times as much," Myhr says. "And the consultant might have recommended a more expensive solution."

HIRING

Choose the Right CIO

Recruiting a chief information officer is tough for company owners who might lack the technical know-how to conduct a productive interview. Placement experts and MIS directors suggest these **key questions**:

Ask the candidate: How does a modem work?

Consider the answer: Is the candidate able to explain simple technical issues without using jargon or assuming a condescending attitude?

Ask the candidate: What is the most difficult technical problem you've solved?

Consider the answer: Was the problem truly difficult? Did she consider the business issues involved?

Ask the candidate: Have you been authorized to make technical purchases? Describe them.

Consider the answer: Can the candidate think through purchasing decisions and identify the relevant business concerns?

Ask the candidate: How would you network our six desktop computers?

Consider the answer: Does she ask insightful questions about our needs and resources? Does she identify how our employees use technology?

Ask the candidate: What do you like and dislike about your current job?

Consider the answer: Does she like having hands-on responsibility for keeping everything in good order? Is she flexible? Will she take on a broad range of responsibilities?

HIRING

Put Candidates to the Test

Exaggerating on a résumé is nothing new, and although some white lies may prove harmless, others can cause painful headaches.

Lori Cohen, manager of training and development for Scott-Levin, in Newtown, Pa., has suffered such headaches. Because the pharmaceutical consulting firm is continually looking for new hires, Cohen had compiled a list of computer terms and functions. She used to ask all new hires to review the list, checking off those skills and terms they knew. She spent hours sorting through those lists by hand, only to discover that the questionnaire had yielded inaccurate results. "A lot of people fudged it or simply couldn't articulate what they did or didn't know how to do," she says.

Frustrated, Cohen scrapped the checklist and turned to a computer-based test she had taken at a previous job. The software, Prove It! (Know It All, 800-935-6694), **forces candidates to demonstrate their abilities rather than just talk about them**. Now Cohen has every person who applies for an administrative or marketing job at Scott-Levin take a crack at it.

She estimates that for every new hire, the software saves her about $550 in training. But the biggest saving has been in recruiting and training replacements. Every unsuccessful hire means a loss of about $5,000. Since implementing Prove It!, Cohen hasn't had to replace a single new hire.

HIRING

You've Gotta Have Heart

When the technology craze began, I was intimidated by machines, so I hired people based on their computer skills alone," explains Samuel Metters of Metters Industries, a 15-year-old engineering firm in McLean, Va. That mistake ended up costing him plenty.

"If a candidate was not computer-literate, and he did not know at least two or three software programs inside and out, I saw no place in the company for that person. I looked for computer nerds and software techies, and loved anyone who could sit in a corner and play on a computer all day.

"As a result, my **employees knew the technical aspects of the business, but they couldn't deal with people**. We started losing a lot of follow-up contracts. I remember presenting a proposal to a government agency and watching my employees ram our ideas down the representative's throat. If the government had to decide between Metters Industries and a company with equal technical resources, the decision makers would choose the other company because we didn't understand how to treat people. I knew things had to change; Metters was turning into a company of robots.

"Now I look for candidates who have that intangible element: a warm side to their personalities."

HIRING

New Hires from All the Right Places

Bill Wood, founder and president of Wood Personnel Service, in Nashville, turned to a geographic information system (GIS) to locate potential hires. Because of the persistently tight labor market around Nashville, Wood Personnel's product—labor—sells itself. But, Wood explains, that in itself is a complication. It's a challenge to find the right people in sufficient quantities. To keep his nine-year-old company growing, Wood assembled a profile of the roughly 500 temps he'd already successfully placed and set out to **recruit workers whose demographic profile** matched that of his current workforce.

How would he locate such employees? Wood found his answer in GIS software that comes with updated U.S. census data. The software maps the data according to color-coded demographic zones. Wood used zip codes to define the areas he wanted to target because, he explains, "zip codes were something I felt I could get my hands on." He booked help-wanted ads in media that served his target zip codes, rented banquet rooms in local hotels to conduct interviews, and recruited from the pool of applicants who showed up. A few months later, he studied the campaign's effectiveness. "The software increased our applicant flow 25%," he reports. And revenue followed "almost exactly."

HIRING

Does It Pay to Relocate?

Offering a job to a prospective employee who would have to move from another city can be tricky. For example, because the cost of living can vary so much from place to place, it's hard to be sure whether the salary you are offering jibes with economic reality. Software called ReloSmart (Right Choice, 800-872-2294) can help you **design a competitive compensation package**. The program's data and calculation tools let you do an apples-to-apples comparison between a prospect's current situation and the one you're offering.

ReloSmart, which was written for employees but is nevertheless useful to employers, first asks the user a series of questions about the candidate's financial position and where he or she is planning to move. Referring to its database of local and state income taxes, real-estate sales and tax information, energy costs, and other statistics for both the new and old locations, ReloSmart calculates the impact of a move. The results are presented in four categories, including Salary—how much is required to maintain the same standard of living in the new location—and Quality of Life—how the weather, crime, median income, and average level of education in the two locations differ. With those data, you're in a much better position to pitch your location to the prospect, and you have the ammunition you need to negotiate reasonable compensation.

HIRING

She *Said* She Was an Expert

"I hired a computer specialist who said she was an expert. She turned out to be the first person I've ever had to fire. That 'expert' practically ruined our computer system." The experience taught Judith Jacobsen how to screen new computer hires. The CEO of a $4-million greeting card company, Madison Park Greetings, in Seattle, tells what went wrong:

"Her previous employer told me she was a good worker. She worked for about three months, telling me that she was making the system easier to use and more versatile. Because I don't know anything about computers, I left her to her own devices. It turned out that she mixed order entries with inventory information and with financials. All the information in our database was one big jumbled mess.

"I first realized that something was wrong when an employee told me that the files seemed mixed up and reports weren't making any sense. I hired a local software-development company to help. The representative took one look and said, 'This system has been sabotaged.' It took them a week and a half to put the system back in order.

"I don't think the woman ruined our system on purpose. I think she was just incompetent. I learned that thoroughly **checking a candidate's background is crucial**. I should have asked her previous employer how good she was with computers instead of what kind of person she was. When I talked to her employer later and mentioned my trouble with her, he said, 'Well, she gave the impression that she knew a lot, but now that I think about it, she was a disaster.'"

HIRING

Ready for a Techie?

Many small companies can't justify hiring a chief information officer (CIO) to manage their information systems. But for a midsize company, the decision may not be so obvious. Answering the following questions can help you **determine when it's time to hire a CIO**:

- *Is your industry information-intensive?* Without a CIO to keep you current with key competitors, even at the low end of the revenue scale, you risk losing market share or reducing profit margins.
- *Are you competing with much larger companies?* If so, you're probably up against some impressive information systems. A good CIO should be able to leverage your technology for competitive advantage.
- *Does your company have a hodgepodge of systems?* A CIO can keep every department and its information technology on the same wavelength.
- *Are your systems unresponsive to business change?* If your systems remain stagnant while your industry changes, a CIO might help your company push the envelope a bit.
- *Do you worry about spending too much on technology?* If your expenditures for technology have climbed to 10% of sales, a CIO can make sure you're getting your money's worth.
- *Are you expanding into international markets?* A CIO can help navigate the tricky border crossings, outmoded phone systems, and Byzantine trade regulations.

Assistants Who Are Never There to Help You

If you know that you need help but still prefer to avoid the hassle of having an employee, try a virtual assistant. Although that may sound like a sci-fi suggestion, a virtual assistant is simply someone who performs such functions as scheduling or correspondence from a remote location. "It's a fast-moving trend," says Charles Grantham, president of the Institute for the Study of Distributed Work, in Walnut Creek, Calif. "We're seeing it pop up all over the place."

Virtual assistants offer several advantages over temporary or part-time workers. First, there are the obvious benefits: no payroll taxes, no workers' comp, no commissions to temp agencies. But also, because they charge by the hour, virtual assistants are generally more cost-efficient. Stacy Brice, a virtual assistant in Cockeysville, Md., charges $50 an hour but discounts that rate to $35 if clients fork over a $525 monthly retainer. The pay arrangement and the distance tend to clarify the relationship, making it less of a management challenge.

"If I have an on-site employee, I've got to be prepared to deal with that person when he or she shows up," says David Goldsmith, a customer-service specialist based in Windermere, Fla. "It's a job, managing an employee." Brice, who assists Goldsmith, handles tasks such as rounding up Japanese business, booking speaking engagements, and identifying Website designers. Relying on fax machines, e-mail, voice mail, and 24-hour business support centers, a virtual assistant can work for clients across the globe.

• MANAGING PEOPLE •

MEETINGS

Don't Erase This Board

As a patent attorney for the Boston law firm of Banner & Witcoff, Charlie Call spent hours in meetings watching his clients scribble technical mumbo-jumbo on whiteboards. The excited company CEOs would struggle to note all their important points, while Call and his colleagues tried to keep up on their legal pads. To make sure he didn't miss anything, Call often had his secretary snap a Polaroid of the board before it was erased, or he'd ask her to spend her lunch hour transcribing and typing the chicken scratchings.

Now, Call relies on an electronic whiteboard called Ibid (MicroTouch, 888-388-4243). He simply hooks the Ibid whiteboard to his computer using a parallel port, and **everything written on the whiteboard is saved automatically** in a computer file. After the meeting, Call can e-mail the information to clients and coworkers, print it out, or cut and paste it into, say, a Microsoft Word document. "I keep the system running all the time," says Call, who gets a kick out of capturing his visitors' doodlings.

Call, who works at home every other day, says the system is great for keeping him up to date on meetings that are held while he's out of the office. And he's not the only one who's thrilled with Ibid. Banner & Witcoff's secretarial staff no longer spends hours trying to decode whiteboard arcana, and Call says, "They really appreciate being able to go to lunch instead."

MEETINGS

Just Listen to Yourself

What can you do to increase employee participation in company meetings? Maybe your employees aren't the problem. Maybe it's you. In particular, maybe it's the way you communicate. Ralph Stayer, CEO and owner of Johnsonville Foods, a sausage manufacturer in Kohler, Wis., discovered the truth when he was trying to encourage employees to take more initiative. Traditional tactics weren't working, so Stayer put simple technology to work. He literally listened to himself talk.

Stayer started to **record his staff meetings**. When, later, he reviewed the tapes, Stayer was surprised to hear himself distinctly stifling employee participation. For example, as soon as an idea was opened for discussion, Stayer himself would jump in with, "What do you think? Here's what I think." Painful as it was to play back his own words, Stayer says the tape helped him understand and change his management style. And employees are now more comfortable voicing their opinions.

• MANAGING PEOPLE •

MEETINGS

Let's Meet: Your Office *and* Mine

Sometimes it makes sense to use **videoconferencing without the video**. "Personal conferencing" is especially well suited to businesses that care more about looking at clients' documents than clients' faces. Here's how personal conferencing works: One computer dials another by modem. The connected parties see the same document, and together, they can mark it up and revise it. Both computers store an electronic version of the working session, a permanent record of the "meeting."

Tucci, Segrete & Rosen (TSR) Consultants, in New York City, uses ProShare (Intel, 800-538-3373) to conduct such online conferences. The $6-million company designs the interiors of department stores for clients like Saks Fifth Avenue and Marshall Field's. Rather than travel to a client's site each time he wants to review changes in a plan he's drafting, president Dominick Segrete conducts online meetings.

The company paid $7,000 for each of several video-ready PCs and their sophisticated software. The videoconferencing software, ProShare, cost a few hundred dollars a copy, but Segrete reports that travel expenses have declined dramatically. The clients, who foot the travel bill, are happy, and meetings themselves are shorter. "When you travel across the country, you feel obligated to make meetings longer," he says. Another unanticipated benefit: Junior employees, who in the past had little opportunity to interact with clients, are sitting in on ProShare meetings. "It's a good way to grow their skills," says Segrete.

MEETINGS

All but Face-to-Face

Rather than conduct on-site meetings for its dozens of field agents in Miami, New York, and San Francisco, Service Intelligence, in Seattle, staged a single two-hour videoconference. Susan Smith, a principal at the $3-million market research firm, says the session cost only $1,500. The same meeting, held the old-fashioned way, would have cost $6,000 in travel expenses, and everyone would have had to set aside at least 40 hours for the event.

Formerly the province of large, capital-rich, or technology-savvy companies, **videoconferencing is now available to companies of any size**. For less than $2,000, a company can equip a PC with the software, video board, camera, and phone lines needed to hold a videoconference with a similarly equipped party. Such vendors as Apple, AT&T, Creative Labs, Intel, and PictureTel sell videoconferencing kits. But you don't have to commit. Like Smith, you can enlist the service of a third-party organizer. Kinko's, for example, offers space and PictureTel equipment for $150 an hour per site.

• MANAGING PEOPLE •

MEETINGS

Online Anonymity Sparks Discussion

Holding electronic meetings can benefit a small company in many ways. Meeting online is cheaper, easier, and as two organizations learned, **e-mail is likely to encourage better discussions**.

Jimmy Carter, former President of the United States and founder of the Carter Center, in Atlanta, explains that during the organization of the Atlanta Project, he and his colleagues held discussions with various community groups, using computers to gather opinions "in an unidentified way." He says, "These were highly revealing, even sometimes shocking to some people on the receiving end—like me. Afterward we did another round on the computers and came up with various opinions on how the problems might be resolved. Then we reached an agreement. The consensus came in just two or three hours; without the computer system it might take weeks of negotiation and repetition. And some of the opinions never would have been raised...some of the folks who worked under me would rarely—if ever—confront me with their gripes. Because the technology allowed them to be anonymous, people were free to say what they wanted without retribution."

At Marketing Partners, a firm in Burlington, Vt., copresident Pat Heffernan used group brainstorming software to hold online sessions with clients. "All contributions were anonymous. The session accelerated the brainstorming by sharpening our concentration and thoughts. The result? An enormous number of ideas in a one-hour session. The anonymity granted by the computer liberated our thinking."

TRAINING

Experience Is the Best Teacher

It's possible to have too much information. Internet junkies find themselves drowning in e-mail. And surfers setting out for a 10-minute ride on the Web are often surprised to discover that three hours have disappeared while they've been searching for an elusive bit of information. Certainly, using the Internet is the best way to learn about its features. But that doesn't mean that your employees should spend hours browsing and searching on their own. **Employees who spend time online can train one another.**

At Jaffe Associates, a marketing firm in Washington, D.C., employees have a simple way to leverage the time they spend online. The firm publishes a weekly internal e-mail newsletter called *Virtual News*. Employees contribute searching tips and review exciting Web sites. Similarly, engineers at Ward & Associates, an environmental-consulting firm in Austin, e-mail one another about great sites they've explored during the week.

Reading magazine articles or books that list popular or impressive Web sites may be informative, but sharing recommendations with one's colleagues is a much smarter way to stay up-to-date. With millions of Web sites appearing and changing all the time, your employees need an effective strategy for getting solid facts fast.

TRAINING

Turn Techies into Trainers

Now's the time to learn to navigate the Internet. The longer you postpone it, the more onerous it will seem.

"You'll be frustrated in the beginning," warns Seena Sharp, principal of Sharp Information Research, in Hermosa Beach, Calif. "You're going to be spending a lot of time figuring it out. You have to learn the shortcuts, and the only way to *really* learn is to do it." Experienced Internet users suggest you **find a few informal consultants to help you learn how to use the Internet**. Look for two kinds of people: technically advanced Internet users and colleagues who are passionate about your area of interest. The techies will save you time by briefing you on the latest advances. Tell them about your interests, and they'll keep you posted on what's new. Your second group of trainers includes people who need to know the same things you do, but who are already comfortable navigating the Net. Keep a list of current concerns, and whenever you get together with your "gurus," steer the conversation to answering your questions.

But you can take your first steps online by yourself: Read a book, take a class, or get a pal to guide you. Deborah Hollander Schwartz is a publicist with Jaffe Associates, in Washington, D.C. Her firm offers marketing and communications consulting to legal professionals. She found the perfect teacher—her 13-year-old son. And she paid him $10 an hour.

TRAINING

Enter Ergonomics

Smart businesses aren't waiting for regulations or lawsuits to prompt action geared to preventing such repetitive-motion injuries as carpal tunnel syndrome.

At Woodpro Cabinetry, in Carbool, Mo., an ergonomics program guided by employee safety teams has reduced workers' compensation costs from $106,000 to $65,000 since 1994. David Carroll, director of safety at the 107-employee company, says the first years of the program saw workers' comp claims rise as employees began identifying injuries as being work-related. But Carroll asserts that it's better to catch the problem "while it's still tendinitis. It may seem counterproductive at first, but full-fledged carpal tunnel syndrome is far more expensive." Carpal tunnel is an injury that commonly strikes computer workers. It's expensive to treat, but if you establish **an ergonomic work space**, you can help prevent it.

By taking advantage of its insurance carrier's prevention programs and training, Woodpro is keeping its ergonomics-program costs down to about $5,000 a year. Employees' suggestions, such as lowering tables and rotating responsibilities, have been "simple and straightforward," says Carroll. "The employees know what feels right."

TRAINING

Back to School—Low Cost and Local

Do some of your employees need training that you fear far exceeds your budget? Don't overlook the resources of your **local community college, which may offer computer courses**, including training on specific software. While services among community colleges vary, a community college is where Bill Carson, president of Cutting Dynamics, in Cleveland, found the courses his employees need—at a price he can afford.

Workers at the $5-million aircraft-parts manufacturer learn specialized skills, like blueprint reading, twice a week. Rather than pay individual student fees for the 18 employees who enrolled, Cutting Dynamics pays for each 30-hour class. Carson expects the class to pay for itself in improved productivity. "Some machine tools cost as much as $600,000," and, he says, he knows that "it costs more if your people don't know how to operate them efficiently."

TRAINING

Start Computer Training at Home

At MDP Construction, in Colorado Springs, Colo., secretary-treasurer Rick Lewandowski had his work cut out for him. The $12-million company was automating project tracking, but MDP's supervisors were all former painters and carpenters who, by and large, were not computer-literate. For some of them, the transition was pretty rocky. Certain employees couldn't get the hang of updating the project-management software when tasks were finished or revised, which naturally wreaked havoc. The time lines bore little or no resemblance to the job at hand. Most of the supervisors did eventually adjust, "but they had to take the initiative and spend a lot of their own time learning the process," says Lewandowski.

To help **familiarize his staff with computers**, he gave the supervisors IBM-compatible desktop PCs for home use. He also provided them with game programs and wrote spreadsheets in Lotus 1-2-3 to help them manage their personal finances. "And I spent a lot of time on the phone, coaching them," he says.

TRAINING

Online Job Training

At Professional Analysis Inc. (PAI), in Oak Ridge, Tenn., managers dreamed of converting the in-house training course into software and installing it on the network. **Employees could then train on their own**, walking themselves through text material and quizzes to make sure they were absorbing the information. Because there was no budget for the project, programmer Ahmad Elhaddad was working on developing the idea whenever he wasn't busy with his official assignments.

Elhaddad started with a relatively simple task: He transferred PAI's 40-page employee benefits handbook to disk. The electronic version not only keeps employees up to date on personnel matters, it also eliminates the need for annual benefits meetings. The disk holds text, pictures, and extensive question-and-answer sections.

The next step is to develop electronic versions of the company's four-part quality-control and OSHA training classes. PAI plans to make this program interactive. For example, when an employee answers a question incorrectly in one of the trial tests, the program will offer the option of jumping back to the appropriate explanation. Human resources manager Jeff Ginsburg is pleased that employees are learning on their own. "It is a much better method of training," he says.

TRAINING

Tell Your Temps Where to Go

Winston Flowers, a Boston florist with five retail outlets, relies on delivery personnel to get the job done. The $14-million company hires extra drivers during peak days, explains owner Ted Winston. "On Valentine's Day we go from 15 drivers to 75." How does Winston guarantee that his temporary workforce can negotiate the notoriously complicated Boston streets to make deliveries on time? He uses routing and mapping **software to provide each driver with a printed map** that shows exactly where each of the day's deliveries is to go. "The routing and mapping software gives the temporary drivers the tools they need to perform like experienced veterans," says Winston.

The software Winston uses, a custom-designed program that cost him $150,000, is an order-entry and tracking program with mapping functions. Winston runs it on the 40 workstations in the company's warehouses and on the 20 terminals located in the stores.

TRAINING

Sitting Prettier

Computer users are a heady—as opposed to corporeal—lot. They invest an inordinate amount of money in computer systems, software, accessories, customized keyboards, and even mouse pads, but they skimp when it comes to a most critical element in a high-functioning office: the chair. It's not unusual to see, in front of a $4,000 computer system sporting all the latest and fastest components, a person perched perilously on a stool. And **an ergonomically deficient work space means skyrocketing workers' compensation injuries.**

Does your computer work space measure up to ergonomic standards?

- *Computer monitor.* The top of the screen should be at or just below the seated person's eye level. There shouldn't be any glare from lighting or windows.
- *Keyboard.* The keyboard should be at the seated person's elbow height. Wrists should be elevated by wrist or keyboard pads. Ergonomic keyboards follow the natural curve of the arms.
- *Telephone.* Clenching a handset between jaw and shoulder is a classic source of neck and shoulder pain. Headsets help prevent strain.
- *Seating.* Invest in a chair designed to encourage correct posture. Poor posture affects the back, arms, legs, and neck.
- *Posture.* It's important to keep moving—a software program like ErgoBreak (Vanity Software Publishing, 800-643-2881) encourages employees to take frequent breaks. A cartoon character leads exercises.

TRAINING

Retain and Retrain

Cook Specialty, a $13-million manufacturer of precision metal parts in Green Lane, Pa., used to take customers' blueprints and knock out a finished product. But president Tom Panzarella realized that if he wanted to maintain a competitive edge, he'd have to overhaul the company's entire manufacturing strategy. He added computer-integrated manufacturing software. But it's not always easy for small companies to go high-tech in the world of manufacturing. Retraining and labor restructuring can be massive expenses. Is it smarter to hire a new staff of experts? Or should you **retrain your workforce**?

When he made the switch, Panzarella chose to retrain his 12 welders. To ease the transition, he searched for and bought the most user-friendly machine available, and he taught the welders how to program and use it. "We'd rather retrain our people to use technology than hire a bunch of textbook-trained experts and try to teach them welding," he says. "After all, our agility depends at least as much on our employees and processes as it does on our machines." Besides that, it's cheaper to teach technology than it is to teach a core competency like welding.

TRAINING

Tips for Training the Technophobic

Gary Mandelbaum of Karman, a clothing manufacturer in Denver, knew he had to install an electronic-ordering system. His sales staff was shipping handwritten orders, and mistakes were costing him customers. But because his industry is traditionally anti-computer, Mandelbaum moved carefully and slowly to ease his sales reps into the information age.

Instead of thrusting the software at the entire sales force, Mandelbaum **selected three reps for a brief pilot program**. To put the software in its best light, he made sure that at least one of those reps had a fair amount of computer know-how. And to build credibility among the most computer-shy of the sales force, he tapped two computer novices to be the other two pilot reps.

The three pilot reps were trained for one week before they set out with their laptops for a high-pressure field test. And all three returned with glowing reports for the other reps. According to Mandelbaum, the strategy was to let the employees "sell one another" rather than relying on management.

The rest of the sales force was quickly equipped with laptops and given four days of training. Mandelbaum outlines the process: "First, we told them only what they needed to know to get their basic functions accomplished to write an order. When they were ready for more, we taught them how to get in and out of the system." The sales force was then turned loose in the field for a sink-or-swim rollout. "We couldn't have gone through another season the old way and survived," he says. Not only did the reps survive, they prospered. Any resistance quickly evaporated.

TRAINING

Phone Home? Only for Business

"My biggest blunder was putting cellular phones in all our delivery vehicles," says Reid Litwack, president of $55-million Action Steel Supply, a steel distributor in Indianapolis. Litwack's problem was that he **handed out phones before creating a company policy for using them**.

"We bought the phones so our truck drivers could keep in touch with the home office without having to pull off the road to make a call. I set things up so the drivers could call only the company and no one else.

"Suddenly, one driver's bills jumped from the company average of $30 a month to $100 a month. I started making inquiries and found out that he was dating our receptionist. Everyone seemed to know about the relationship but me. At first he hemmed and hawed, telling me that he just happened to have made a lot more calls to the office than most of the drivers.

"I told him that from then on he'd have to pay for any amount billed to his phone over the average $30 a month. Surprise. The next month his bill fell right in line with the $30 average. There have been other romances in the company, but they didn't cost me any money."

RETENTION

How to Keep Your Techies Happy

With unemployment at remarkably low levels, finding skilled technical workers is difficult. And retaining talented employees is even more of a challenge. Here are two tips that can help **keep your technical employees from returning headhunters' phone calls**:

- *Offer employees a piece of the action.* To keep the programming staff he's assembled, Raoul Socher of Network Software Associates, in Arlington, Va., pays his software developers royalties on the products they help produce. The catch? Programmers are eligible for the royalties only as long as they stay with the company.

- *Give them time off.* Daniel Maude is CEO of Beacon Application Services Corp., a $10-million software-services company in Framingham, Mass. He shortened his software consultants' workweek. Because Beacon's consultants do most of their work at distant customer sites, they had to leave town on Sunday and couldn't return until Friday after work. "That left them with only one day off," says Maude. Now those employees leave Monday morning and return Thursday night, still working roughly the same hours. "If you want to keep people 5 to 10 years, you can't expect them to work 60 hours a week," Maude says.

WORK SPACES

Bringing It All Back Home

The thought of paying rent may make you want to ditch the idea of a corporate office altogether. That's what Janet Caswell did. As Caswell pondered space options for her accounting firm, Caswell & Associates, in Bloomfield Hills, Mich., she thought about the staffing and technology changes that were transforming her $650,000 firm. More employees were asking for flexible schedules, and clients were looking for specialized knowledge that Caswell could get most efficiently from part-time outsiders. She realized that she "didn't need to spend $22 per square foot just to have file cabinets."

So Caswell quite literally **sent her people home**. She gave up her office space and set up each employee—herself included—with a home office. Caswell spent from $200 to $500 for each worker's modems and extra phone lines, and from $2,000 to $3,000 for each employee's computer and fax machine.

Since the transition, Caswell's phone bills have doubled. But all the costs together don't begin to approach the more than $3,000 she would be paying in rent each month.

WORK SPACES

Offices without Walls

Innovations in the work environment typically come from professional service firms. These two small companies established **unusual work environments** that work.

In 1996, Cambridge Management Consulting, a division of Cambridge Technology Partners, moved into a restored warehouse in San Francisco. Consultants walk into an open space dotted with unassigned cubicles. They may choose to work in cubicles, in a team room, in the lounge area, or even under the basketball hoop that hovers 10 feet above the ground of the exterior courtyard. "Our company culture is manifested in the layout of the office. It includes workspaces and team rooms that capture both our entrepreneurial spirit and team-driven environment," says Chris Greendale, the company's acting president.

The programmers at SHL-Aspen Tree Software, in Laramie, Wyo., need plenty of opportunities to interact with one another because producing software is a team effort. The $7-million company has helped such companies as American Express and Marriott computerize employee interviews. To encourage interaction, company president Brooks Mitchell gutted two houses, set up servers in the basements, and told his 35 employees to plug in their laptops anywhere. "I want my company to feel like a family, and a family lives in a house," he says.

WORK SPACES

Workers Who Stay Put

Before you begin thinking about creating an alternative work space, think about your employees and their responsibilities. Some workers don't fit comfortably into nonterritorial work spaces, and sometimes a company's operational processes don't lend themselves to such an experiment. That's what Charles Rodgers learned at Work/Family Directions, a $65-million consulting agency in Boston.

Rodgers tried a pilot program in which the company's 34 telephone counselors worked in 6,000 square feet of unassigned space. The area was designed to allow counselors to move among a number of cubicles equipped with computers and phones.

But there was a problem: The switchboard had no efficient way to route incoming calls to the counselors, who were always on the move. "We learned that we need to do more **reality testing before implementing a radical work space change**," says Rodgers. For the moment, the office is back in a more traditional arrangement.

WORK SPACES

Wide-Open Spaces

At AGI's Melrose Park, Ill., headquarters, there are **no traditional private offices**. Instead, workers at the packaging manufacturer operate in an open environment with few walls and doors. No matter where they are—at a coworker's desk or on the plant floor—they have access to their computer files from any terminal or PC on the premises. Those who use laptops can plug into the company's network at any of 250 data ports. Many employees carry pagers because they spend 80% or more of their time away from their desks, solving problems on the fly.

CEO Richard Block invested nearly $1 million in the nonterritorial work environment when AGI sales were at $30 million. The offices are in a square with an oval track in the middle. Outside the oval is open space filled with modular office furniture; there are no walls, and the area is flooded with sunlight from skylights. Inside the oval's perimeter are executive offices with glass walls and no doors. And at its very center are conference rooms.

Ad hoc teams are the rule, and Block spends as much as 75% of his time walking around, talking to designers, pressmen, and clients.

WORK SPACES

How to Kick the Paper Habit

"I had to do something dramatic or people would have gone on using paper forever." That's how Dan Caulfield, founder of Hire Quality, a job-placement firm in Chicago, justifies his rather aggressive approach for **converting his company into a paperless office**. One morning, he stormed through offices, snatching paper from employees' desks and burning his collections in a trash barrel on the fire escape. Of course, Caulfield had put systems in place to facilitate an orderly transformation.

In contrast to the $5 to $7 it used to cost the company to register new candidates, today it costs only $1.50. Here's what Caulfield did:

- He provided employees with a centralized database containing all the information they need.
- He installed scanners to convert incoming paper documents to computer files.
- Employees who use the least amount of paper are rewarded, and those who use the most are penalized. In the early days, Caulfield used to put a jar next to the printer to collect fines of $1 for each fax and 25¢ for each printed page.
- He encouraged clients to use e-mail rather than paper faxes.
- He offered clients incentives to implement automated delivery systems, accepted electronic invoices, and made electronic payments.

WORK SPACES

Strike an Office Deal

Overhead can eat up a company's bottom line, and one of the fattest overhead expenses is office space. Some companies avoid this expense by establishing virtual offices. Employees work in remote locations (usually at home) and use telephones and modems to connect to a small central location.

For companies that require true office space, there may be another way to save money. Companies that are smart about leveraging technology can **swap technological expertise for office space** supplied by a client or a vendor.

When Paul Upton's Precision Computer Service, in Oklahoma City, outgrew its first office, Upton turned to a customer with space to spare. The client had hired Precision Computer Service, now a $16-million company, to repair and service computer systems. As part of the maintenance contract, Upton proposed a 5% discount in return for office space. Not a bad deal even in 1986. Upton's monthly rent amounted to a piddling $125. The steal of a deal lasted for about 18 months, ending only when the client closed its Oklahoma branch.

WORK SPACES

Low-Tech Touch

Sometimes new equipment is not the answer to increasing productivity, and simply rearranging the equipment you already have can make all the difference. Ken Rizner, vice-president of manufacturing at Hyde Tools, in Southbridge, Mass., walked through the steps of every process with teams of line workers. When they really understood all the steps it took make each hand tool, they were able to eliminate the steps that added no value to the final product. And they were able to **rearrange machines and toss arcane reports**.

The combination of two major operations cut out at least 30 steps, and Rizner estimates that Hyde sliced some nine-tenths of a mile from the production process. The new systems reduced the time—from more than 10 weeks to 15 working days—it took to turn raw material into finished product. Imagine getting the same results in your order processing or billing departments.

WORK SPACES

Store Your Records Electronically

The problem, everyone thought, was that there wasn't enough space for ingredient storage. Producers Dairy Foods, a manufacturer of dairy products in Fresno, Calif., was quickly filling up the 10,000 square feet it had allotted for document storage. The files and documents were practically spilling into the warehouse space designated for ingredients. But, according to chief financial officer Frank Sewill, the real problem was too much paper. Producers Dairy was keeping hard copy of all its financial records, customer invoices, sales reports, and product information.

Sewill recognized that the company had to find an alternative to paper records. He purchased a DataView **laser-disk storage system** (MultiProcess Computer Corp., 800-377-5681) to replace the shelves of paper. Now, Producers Dairy stores all its documents—from financial reports to customer invoices—on laser disks. Records that used to take up nearly 10,000 square feet of storage space now fill an area the size of two shoe boxes.

Employees were worried that they would lose important documents, but they soon realized that it is much easier to call up a computer file than it is to search through boxes of documents. Sewill expected to save time and space, and he was more than pleasantly surprised to discover that he was also saving some $40,000 a year in paper. Paper consumption declined from nine boxes to less than one box a day.

IV

"While one wants to avoid the bleeding edge of technology adoption, you don't want to go down the other extreme and keep waiting for prices to go down. If you wait two years, you will enjoy twice the value. That is true. But the cost of the wait to your company may be much higher."

NICHOLAS NEGROPONTE
Professor of Media Technology, MIT,
Cambridge, Mass.

BANKING

Banking Goes Virtual

Every afternoon, Allen Systems Group's chief financial officer, Frederick Roberts, sits down with the company's Miami banker. But there's a twist—Miami is 100 miles away from the $26-million international computer company's headquarters, in Naples, Fla., and Roberts **does all of his banking on his PC**.

Many bankers are thrilled at this new wave of cyberbanking. "We want our customers never, ever to need to walk into a branch office of a bank again, unless they want to," notes Thomas Hoffman, Bank Leumi USA's first vice-president and head of custom business services. "If they've got PCs, they should be able to process their payrolls, reconcile their accounts, and switch excess funds between investments without leaving their offices or calling a banker."

That's certainly the way things work at Allen Systems these days. On a typical morning, Roberts and his staff pore over faxes that detail the company's international accounts in Europe and the Far East. By 2 p.m. they've compiled a daily treasury report. CEO Art Allen and Roberts then meet to decide how to lower the company's expenditures and enhance its cash flow by moving funds among its various domestic and international accounts.

"There are so few small companies out there that even know about this," says Allen, "that it's hard not to view this as our competitive edge. Boosting operating cash flow through more effective cash management reduces our company's need to go out and borrow those funds."

BANKING

Why You Oughta Automate

It was the bank that forced Clarklift/FIT, in Orlando, Fla., to invest in computerization, and that probably saved its life.

When the company's president first began talking to Citicorp about capital to fund his three forklift dealerships, the bank explained that it would require the company to track its own inventory and submit monthly status reports. But there was a sticking point: Management would first have to prove that the company's computer systems were up to snuff. Clarklift/FIT found itself in the middle of a trend: **Banks want their small-company investments to have solid computer capabilities**.

"It's simple," explains Citicorp's David Hilton, vice-president for the Global Equipment Finance Division. "If clients don't have the proper systems to automatically generate the sort of financial reports that we need on a monthly basis, this sort of loan would overwhelm them."

Citicorp sent an independent assessor to audit the company's information systems. By that time, management had invested some $100,000 in hardware and a proprietary inventory and accounting system. For an entire week, the assessor hit CFO Ken Daley with a barrage of financial questions that had to be answered quickly by culling data from the computer. In the course of putting the computer system through its paces, Daley uncovered and discontinued bad business practices that had become institutionalized, and several months later, the bank agreed to give the company $5 million in credit. After pouring about $400,000 into technology over two years (about 1% of company revenue during that period), net sales grew by 15%, and cash flow improved by 14.5%.

BANKING

Save Cash: Bank by Computer

If you've been overwhelmed by all the hype and still aren't clear about what high-tech banking means, here's an overview. The biggest game in town is **electronic cash management**—the ability to access account information directly, without a visit to a real or an automatic teller. The most technologically advanced banks allow their customers to use their computers to dial in for up-to-the-minute accountings. Others can provide customers with account-balance updates at specified daily intervals, either by fax or secure voice mail.

How does bank automation benefit a business? With timely information, business owners can make better decisions about how to manage their funds. Online banking lets them move cash among accounts, paying bills while still earning as much interest as possible on their cash reserves. With PC banking, cash-management decisions are transmitted from business to bank by computer. These days, companies can use their computers to control automatic payroll plans as well, and they no longer have to phone or fax salary-payment instructions to their banks. And the computers can also handle tax-payment authorizations and international currency maneuvers.

Because banks save money by reducing their use of paper, it may pay to ask your bank to give you a financial incentive for putting your company online.

BILLING

Click On to Instant Credit Checks

In 1995, when Richard Pollock bought International Neon Products (INP), in Chicago, he knew that the sign-supply distributor had lost several hundred thousand dollars to bad debt. Still, Pollock was confident that technology could save the 50-year-old company.

Bad-debt and collection issues were starving INP, but with 11 employees and $3.5 million in revenues, the company was too small to have a full-time credit manager. It was left to the accounting staff to call references and banks for background checks on potential customers.

Gathering credit information on public companies is fairly easy, but what about the small, privately owned businesses that comprise the bulk of INP's customers? Pollock discovered an inexpensive **online service that provides credit information about small businesses**.

Here's how the service, Risk Assessment Manager (RAM) from Dun & Bradstreet, works for INP. In addition to paying $5,000 for a workstation, the company pays $3,250 annually to track its 500 accounts. Pollock uses a modem to access RAM, which, in minutes, performs due diligence on thousands of accounts and approves, denies, or limits credit, depending on risk factors. RAM also monitors high-risk accounts by tracking their credit scores.

"The real bonus," notes Pollock, "is that after an account is established, RAM keeps track and tells me whether my customer has gotten stronger or weaker." Pollock estimates it costs him only $20 each time he checks an account. "It really pays for itself," he says, crediting RAM with bringing INP into the black in less than two years and enabling it to keep debt "well under" 0.2% of sales.

BILLING

In Today's Dollars, Your Bill Comes to…

Six foreign languages and 13 currencies made overseas invoicing a headache for Cannondale, in Georgetown, Conn. The bicycle manufacturer, which does about 46% of its sales in Europe, adapted off-the-shelf software to produce invoices in each customer's currency and language.

The **translation software simplifies billing** of Cannondale's European distributors and retailers, making it easier for the U.S. company to blend in. "We're transparent to the local customer because we can invoice in the local language and currency," says Dan Alloway, Cannondale's vice-president of sales and European operations.

Programmer Peg Beasley spent about three weeks rewriting the invoicing and billing functions of the business-planning software that Cannondale bought from SSA, a software-package application vendor in Chicago. At first, Beasley organized files by country, then, when she remembered that several languages are spoken in more than one country, she switched to a language-based system.

Using Beasley's system, when a sales rep enters an order, it is sent to Cannondale Europe's central office in Holland, where invoices are generated based on the customer's "language code." The system ensures accurate billing by accounting for fluctuating currencies. Once a month, Cannondale's bank provides current exchange rates for the countries where it does business. As employees key the figures into the system, it recalculates accounts receivable. That's a helpful feature, whether you're booking $1 million in overseas sales or, like Cannondale, $85 million.

BILLING

But the Savings Aren't Under the Mattress

Michael Bryant, CEO and sole employee of Career Transition Services (CTS), in Baltimore, believes that for a business his size, **financial software would be a waste of time**. He isn't a Luddite. He just relies on computers for more cost-effective applications.

For bookkeeping, he tracks names and addresses the way our grandmothers kept recipes: on three-by-five cards, stowed in little plastic boxes. Whenever he has a business expense, he puts the receipts in an envelope marked, say, "CTS Parking" or "CTS Supplies." At the end of the year he adds up all the receipts.

As for billing, most clients who come to Bryant's office write a check before they leave. But if he has to mail an invoice, he puts a copy in a folder with the client's name on it. When the check arrives, he marks the invoice and puts it into another folder containing all the other paid invoices.

"That thing works like a charm," he says proudly, pointing at the shoe box full of envelopes. "I'm not spending all those hours entering things into a computer, either, just so it can spit something out one time."

BOOKKEEPING

Do Your Own Taxes

David DeLong, chief financial officer at Dynamix Group, in Roswell, Ga., was going nuts. Spread before him was a daunting array of corporate tax forms and pages of complex tax regulations. DeLong knew that his strong background in finance wasn't enough to tackle the tax implications of the company's incorporation and grand opening a week later. But he also knew that Dynamix, a computer remarketer, still wasn't ready to hire an accountant: The company had yet to book revenues, and DeLong was determined not to hire an accountant simply to do his taxes.

He turned, instead, to TurboTax for Business (Intuit, 800-4-INTUIT), one of the most popular **consumer business-tax preparation software** programs available. "It was like sitting across the desk from an accountant asking questions," DeLong says. "It took a few hours. But it probably went more smoothly than speaking to a person because the software was methodical, and we didn't get sidetracked."

TurboTax for Business won't replace an accountant. It doesn't provide the wide-ranging financial management and planning services that your business may need. But if you can go it alone and want to do your taxes accurately and efficiently, or if you do need an accountant but want to save on tax-preparation fees, software is the way to go. DeLong notes approvingly, "You never even have to look at a form."

BOOKKEEPING

Run the Numbers at Home

What happens to a small company's books when the owner takes extended time off to have a baby or for medical reasons? Amy Miller, CEO of Amy's Ice Creams, a $2.3-million chain of seven ice-cream stores headquartered in Austin, Tex., explains how she **handled her company's financials from home**:

"My office manager and I both had babies a few years ago, and I didn't know how we would manage. I discovered that technology has given us the edge the larger companies have as a matter of course.

"With my Macintosh PowerBook and an ink-jet printer, I did everything at home. Using Quicken, I handled all of my company's accounts-payable transactions. I also modemed the payroll information directly to my payroll company. When we applied for a loan from the Small Business Administration, I was able to compile all the historic information I needed for the application without going into the office. We got the loan—$450,000. I even designed a new bonus plan at home for our managers, using the fax-modem on my computer to send and receive information from the office. I went into work only about once a week, to pick up a stack of bills."

BOOKKEEPING

Accounting by E-mail

Is your business too small to afford top-quality accounting help? Shop around for a **virtual accounting relationship**. Chief financial officer Brendan Burns says that a primarily electronic relationship has given AdOne Classified Network, in New York City, the accounting expertise it needs at a time when the company can't afford and doesn't need a full-time controller.

Here's how AdOne, a $1-million provider of Internet classified ads, works with its accounting firm. "We receive and generate lots of invoices and other financial documents," Burns explains. "Weekly, we send a package to Virtual Growth, our CPA firm, which inputs the data into a QuickBooks computer file. Virtual Growth then transfers the information to us electronically, all properly classified. We update it ourselves and produce accurate, timely financials each month."

AdOne communicates with Virtual Growth mostly by phone or e-mail. Burns is satisfied with the setup. "I won't hire someone internally until our accounting business grows to about 20 hours a week of work," he says.

BOOKKEEPING

Accountants Online

Small-business owners often hate their accountants: They're either too big to pay enough attention to them or too small to serve all their needs," Stephen King says. King, a CPA refugee from Ernst & Young, started his own accounting firm, Virtual Growth, in New York City. His high-tech approach to accounting may give you some ideas about how to structure a workable deal with your accountant.

King assembled a network of small local accountants, who together form a sizable data center. The network offers clients a range of specialized services. Here's how it works:

A small-business owner goes to his or her local accountant. When the accounting needs of the business surpass the expertise of that local practitioner, the owner or accountant, using computer or telephone, outsources that work to Virtual Growth. For business clients who don't need on-site hand-holding, all accounting needs can be serviced by specialists from the Virtual Growth network.

King also uses the Internet so he can act as CFO for his clients. In such cases, clients use their computers to transmit weekly or biweekly financial results—sales, invoices, and the like—to Virtual Growth's headquarters. "And," King explains, "they can download anything they want from their data file anytime. They have **access to financial information 24 hours a day**. That's much more than most other growing companies can say."

BOOKKEEPING

Excellent Service from External Accountants

Gina Slater Parker, CEO of $1-million Hill Slater, in Great Neck, N.Y., appreciates the value of **outsourcing her accounting department**. Several years ago, she hired the Mastermind Group, a bookkeeping and accounting service in Huntington, N.Y., to take over all the accounting functions for her 17-employee engineering and architectural-support firm.

"I wasn't getting the service I needed from my accountant," she says. "No managerial reports, nothing." Since 1984, when she started Hill Slater, Parker had gone through several accountants and bookkeepers. But when she discovered Mastermind, she recognized an opportunity to help her business, and sales climbed 30% the first year.

Similarly, Statistics Collaborative, in Washington, D.C., is a $500,000 consulting firm that turned all accounting functions over to BusinessMatters, in Silver Spring, Md. That business-services firm takes care of everything, from paying electric bills to collecting accounts receivable. Statistics Collaborative president Janet Wittes relies on the reports she gets from BusinessMatters. Each month, she receives a 50- to 100-page breakdown of every aspect of Statistics Collaborative's operations. The series of ratios included in that report track everything from the productivity of each employee to the profitability of each job.

"One monthly report made clear that I had to fire someone," recalls Wittes. A programmer had a relatively low productivity number. At first, Wittes attributed this to the fact that the programmer was new. After a few months, however, nothing had changed. Another employee, Wittes discovered, was losing money on projects. Further probing revealed that she wasn't redefining her budgets when she incurred unexpected expenses. Sharing the information with the employee led to a complete turnaround.

BOOKKEEPING

Choose the Right Accounting Software

Not all accounting software is created equal, and you shouldn't pay extra for useless frills. Here are some shopping guidelines:

If your needs are simple—issuing invoices, writing checks, and printing basic financial reports—look for a **basic accounting program** like One-Write Plus (ADP, 800-649-1720). The data-entry forms look like the forms that you fill out by hand when you use a manual one-write check system. The software tracks salespeople's figures and handles payroll, but if you need to monitor inventory or bid a job, this isn't the program for you.

To track sales information, look for something like Profit (Champion Business Systems, 800-243-2626), which generates two kinds of invoices. A service invoice shows the date and service rate for each line item, and a product invoice shows ID number, description, price, and quantity for each line item. You can record several types of transactions, and each line item is taxed individually.

If you want job-costing capabilities, most accounting software can track project costs, but a sophisticated program like Peachtree Complete Accounting for Windows (Peachtree Software, 800-228-0068) lets you break each job into such mini-components as the job itself, phases, cost codes, and cost types.

Tracking inventory or retail goods can be handled by several low-end programs like M.Y.O.B. Accounting (BestWare, 800-322-MYOB). Look for features that handle parts assembly and automatic back orders. Also, whether your company uses accrual or cash accounting, make sure the software can handle the system you use.

BOOKKEEPING

Timing Is Everything

You've found the perfect accounting software program, the staff is excited about using it, and you can't wait to fire it up and start generating reports. But wait a minute! Before you install the system, think about **when to automate the books**.

The timing of the installation can make or break implementation, especially if there are snags during the process. Good accounting software tracks sales, customers, vendors, receivables, and payables. Just imagine if all your data were thrown out of whack during a peak-sales period. Check your calendar and last year's sales figures to gauge installation times. Don't start the installation if you're about to enter, or are already in, your busy season. You certainly want to avoid inaugurating such a project when your key staff members are on vacation, and you will want to consider the start of your fiscal year.

BOOKKEEPING

Choosing Software? Ask Around

Most large corporations rely on customized accounting software that does everything from producing data for financial statements to calculating payroll for thousands of employees. A $100,000 program is likely more than most small businesses need. But there is plenty of ground between expensive custom programs and juggling scraps of paper. Today's "low-end" accounting software functions like high-end programs, but generally costs less than $400. Here are some tips for **choosing an inexpensive accounting software package**:

- Identify your needs. Consider more than just the accounting aspects of the programs. Adopt a management point of view. Take a good look at how your business functions and decide which features will keep things running smoothly.
- Consider the users' skills and eagerness to learn a new program. What kind of training are users going to need? Have they worked with an accounting program in the past? Do they like working with computers? Users should be part of the software decision.
- Be sure your hardware meets the software's requirements. Accounting programs use a lot of hard-disk space, anywhere from 3 MB to 38 MB, and you'll need plenty of room for your data, too.
- Talk to people who have first-hand experience with the program you're considering. Ask industry colleagues what they use. But just because certain software didn't work out for one company is not enough reason to cross it off your list. Find out what the problem was. Perhaps the difficulty was more with the training or user than with the program itself. Call the software company directly.

BOOKKEEPING

Software for Tracking Costs

Mark Persitz, CPA and owner of Persitz and Co., in Farmington, Mich., needed a way to help his client. The client, a church, wanted him to track rental use of its hall for weddings and other occasions. The church's bookkeeper had been using general-ledger software that required creating a new account to record income and expenses for each event. By year's end, the general ledger was cluttered with one-time-only accounts.

With job-costing software, Persitz was able to **post event information to the same general-ledger accounts**. Because Persitz had used job-costing software before to track material and labor costs for contractor clients, he saw that it could also be used to track such function costs as caterers, bakers, cleaners, and linen suppliers.

He also discovered that the software gave the church an easy way to compare vendor costs, ensuring that events stay within budgets. Now, having a much clearer picture of costs, church officials are considering revising rental rates.

COST CONTROL

Pinpointing the *True* Costs of Sales

"Arbitrary allocation of overhead totally confuses the cost picture of a product," says professor Bala Balachandran, director of the Accounting Research Center at Northwestern University's Kellogg Graduate School of Management. "I have seen cases where companies killed profitable products and increased production of unprofitable ones because they relied on inaccurate data."

Some companies seem to give up on overhead allocation. Their financial statements display a single, outsized line item for selling, general, and administrative expenses. "It's crazy," Balachandran says. "There are **different costs associated with different customers and products**. Why should you lump all of them together in a meaningless whole?"

Activity-based cost accounting (ABC) software was created to allocate costs based on activities performed. Broadly described, the process has three steps:

1. Define such cost categories as salaries, raw materials, travel, and utilities.
2. Identify key processes and the principal activities associated with each; determine activity costs.
3. Assign costs to appropriate categories, like products or customers.

Qtron Inc. applied ABC software, and now management knows exactly what each product's costs are. The San Diego-based company provides contract manufacturing services. Overhead is 80% to 90% in Qtron's industry, and allocation errors lead to pricing errors, which could easily bankrupt the company. Qtron's bids specify all activities—parts ordering, moving, inspections—along with the costs of those activities. Within five months of shifting to the ABC method, the company had enough data to clarify its product-cost structures.

COST CONTROL

The ABCs of Cost Analysis

Management at Koehler Mfg. experienced a rude awakening—its best-selling product line, lead acid batteries, was devouring company profits. The Marlboro, Mass., company had dutifully accounted for the costs of processing and disposing of the hazardous by-products of battery manufacturing. Even with those expenses, the books showed that the batteries were emphatically profitable. The books, however, were wrong.

Koehler's confusion is far from unique. It's quite possible that your most profitable product is actually a loser and that your best customers are costing you much more than they're worth. Does your company sell more than a dozen or so products? Are annual sales more than $3 million? Are your customers' buying patterns wildly divergent? If so, you may not know your costs as well as you think you do.

Until Koehler installed NetProphet (Sapling Corp., 888-335-5051), an **activity-based cost-accounting** package, nobody knew that certain administrative costs were attributable to the battery business. When that information emerged, the product's profitability dropped by nearly 30%. The company hadn't considered product-specific expenses incurred while it was dealing with environmental officials, applying for permits, and filing compliance reports.

COST CONTROL

When Talk Is Cheap

When you have a one-office business, buzzing a coworker down the hall or sending an interoffice e-mail doesn't cost much. But things can get out of control when "down the hall" is 500 miles away. Brock Berry discovered that when the phone bill for a dial-up modem connection between company headquarters in Dover, Del., and its branch offices shot up to $1,400 a month. And that doesn't count the $150 for phone and fax. "When you go online, the meter is always ticking," says Berry.

Berry, vice-president of Berry Van Lines, a $10-million moving company, realized he could lower his phone bill by **connecting the offices with a leased line from the phone company**. Once attached, a dedicated line allows users to send e-mail, voice mail, and faxes for a fixed monthly fee.

Berry installed a 56-kilobit line for $1,200 a month using a telecommunications provider, and now, when he wants to phone a coworker in the company's Baltimore office, Berry just hits a button on the phone. The entire setup cost about $8,000, but Berry figures it was worth it. He's saving at least 20% on his monthly phone bill.

And then there's the nifty service he offers selected customers of the Baltimore and Wilmington, Del., branches: phone numbers that connect to the Dover office. When a customer dials the number, the local phone system transmits the signal to Berry's phone. The customer gets to make a toll-free call, and Berry doesn't have the cost of maintaining a toll-free line.

COST CONTROL

Watch Your Postage Meters

When Federal Express lowered ABL Electronics' shipping prices, the cable manufacturer started billing clients according to the new rates. But FedEx had neglected to update the Powership machine that it lent ABL as part of its service. Owner Randy Amon never checked daily invoices against what ABL billed its customers. For seven months, Amon says, "we were charging customers $9.50 to ship a box that FedEx charged us $10 for." Then Amon noticed the **startling difference between his freight charges and freight costs**. A quick check with FedEx revealed the problem.

"I was blindly trusting technology," Amon says. "Now I'm almost technophobic." Each day, ABL's payables clerk spot-checks the pricing of 5 to 10 shipments. The most common errors are shipments that ABL forgot to bill for. "That can add up to $200 a day," Amon says.

The Hunt Valley, Md., company eventually received nearly $11,000 in refunds from FedEx's oversight. Now that the company does so much international shipping, Amon is really on his toes. He negotiated still lower rates and regularly averages $2,000 a week in refunds when, by using Powership's compliance-reports software, he finds FedEx delivered packages after its promised deadlines.

COST CONTROL

Time Still Equals Money

Joe Phelps is CEO of the Phelps Group, a $34-million marketing communications agency in Santa Monica, Calif. The 54-employee firm has grown at about 25% annually for the past nine years, during which Phelps has learned a thing or two about communication.

Phelps stresses the need to tap the appropriate tool—e-mail, voice mail, or "facemail"—for each task. "When you walk into my workstation to give me a little schedule change, you've used the wrong medium. E-mail me with that. You don't need a lot of bandwidth. On the other hand, if you try to handle an emotionally charged subject with e-mail, you're using the wrong medium, too," he says. But **each medium has its price**.

For example, everyone at the Phelps Group can send e-mail to the entire company. Not that long ago, a hungry employee e-mailed the entire company: "Who took my slice of pizza?" Armed with a perfect teaching tool, Phelps called a meeting of the entire company. Assume, he instructed his employees, for mathematical simplicity, a billing rate of $60 an hour. Say it took each of 50 employees half a minute to open the pizza message, read it, and put it in the trash. Twenty-five minutes at a dollar a minute. That adds up to $25 to try to finger a pizza thief.

"We all share 40% of the profits, so everyone quickly grasped what general e-mail distribution means," says Phelps.

REAL WORLD

• MANAGING MONEY •

"Looking for a trend? Consider selling and buying through online auctions. In time, online auctions may do more to bring order to Internet commerce and lower the costs of bidding. They may actually end up altering our basic pricing mechanisms."

PIERRE OMIDYAR
creator of AuctionWeb software
and CEO, eBay, San Jose, Calif.

COST CONTROL

Cost Cutting Goes Electronic

Many professional service firms earn the lion's share of their revenues from a handful of clients, and they use technology to add depth to customer service.

That's what architectural-lighting firm Fisher Marantz Renfro Stone did. The 20-employee company was riding high during the 1980s, but when budget axes fell in the 1990s, the $3-million New York City firm found that it had to cut its fees by as much as 25% to compete with low-cost one- and two-person shops.

Fisher Marantz persuaded some of its clients to conduct business **online to control meeting and travel costs** and keep service quality high. The company started with e-mail and, as it grew more confident about the technology, began sending electronic blueprints to customers' computers. Transmitting a blueprint electronically costs 35¢ and takes 20 seconds. Compare that to $25 and 24 hours for overnight delivery.

CASH FLOW

A Daily Dose of Numbers

How often should you track your numbers? Daily, Ron Friedman insists. The CEO of Stonefield Josephson Inc., an accounting firm in Santa Monica, Calif., says, "Every morning by 9:30, I receive a printed report that tracks certain key results from the day before. That's a tremendous management advantage. I can respond immediately to any problem signals. Think of all the time and money you lose when you find out about problems only at the end of the week or the month."

Friedman is convinced that watching numbers daily is as important for his clients as it is for his own 75-person accounting firm. "Depending on the type of business you're in, the numbers you need to watch this closely will be different," he explains. "Key numbers might be how much was sold each day, how much was shipped, how big your backlog is, and how much was collected." To make certain that daily reports are user-friendly, Friedman keeps them short. And he knows how **numbers should compare with daily target results**. "Small fluctuations are only natural, but once you track daily results for a while, you'll get a feel for those fluctuations that are more troubling," he notes.

CASH FLOW

The Rapid-Refund Expense Report

If a cash report or a calendar can be sent out via e-mail, why can't an expense report be sent the same way? Most employees are well acquainted with database and spreadsheet applications. At Collectech Systems, a $12-million collection agency in Calabasas, Calif., Chris Murphy, a regional manager, tinkered around with Excel (Microsoft, 800-426-9400) and created a **simple electronic expense form** that everyone now files by e-mail.

Murphy's form, which is just a spreadsheet wearing a little makeup, handles all the computations automatically and goes by e-mail directly to the right people. It shaved two weeks off the reimbursement process. "The reports are legible, so they reduce error and take less time to check," says Britt Johnson, the accountant who cuts the checks. "We can turn one around in a few days." What about snafus or lost reports? "We have fewer problems since people started filing them electronically."

"At the company where I worked before this one," recalls Murphy, "you'd want to kill yourself, waiting for those checks. It could take six to eight weeks. But here the company is not using the float on its employees," he says. "And that makes a difference to us."

CASH FLOW

Just-in-Time Financials

In June 1996, Ray Finch made a last-minute decision. The president of Emerald Dunes, a golf course in West Palm Beach, Fla., decided to run newspaper ads and local TV spots during the U.S. Open Golf Tournament. He wanted to promote special off-season summer rates.

While studying his computer-generated **daily revenue reports**—a must-have feature that Finch had demanded when the developer was writing the industry-specific financial software program—he had noticed a softening of business. If, like most operators, Finch had waited to read a monthly report, he'd have missed the U.S. Open TV audience and gotten a later start on publicizing his aggressive pricing. It was an example of just-in-time marketing that boosted rounds by 28% the following week.

Tee times are like airline seats—you fill them or lose them. "Our inventory is actually little chunks of time," says Finch, explaining that he can easily lose up to 10% of that inventory by failing to note a busy day in need of double-teeing, that is, starting foursomes off holes 1 and 10 simultaneously. "Our software tells me how best to utilize our time."

CASH FLOW

On the Road and on Top of the Numbers

Even managers at the smallest businesses benefit from keeping on top of such daily numbers as sales, expenses, inventory, and employee sales performance. Although that can be tough for the CEO who is also the company's top salesperson, staying in touch with business statistics can make all the difference in the world.

Here's one solution: Have your **alphanumeric pager send the key numbers** to you when you need them. When his company, Computer Gallery, in Palm Desert, Calif., was smaller, Joseph Popper used to set his Excel (Microsoft, 800-426-9400) spreadsheet to e-mail a limited list of key numbers—receivables, sales, and bank balance—to his pager. He'd get the updates several times a day, no matter where he was. "If you grow too fast, you can run out of cash, even with 30-day receivables," Popper warns. "We've been undercapitalized from the start. This way, I know where we are all the time."

CASH FLOW

Job-by-Job Expenses

Can your accounting system produce a report for a particular project—an advertising campaign, say, or a new fall clothing line—that gives you a **detailed picture of overall cash flow**?

If you're using only the accounts-payable and receivable modules of your accounting package, you may have to calculate those figures yourself. That's what the bookkeeper at Hoodoo Ski Area, in Sisters, Ore., used to do to keep track of the 50 or so campgrounds that Hoodoo maintains every summer. Then, determined to find a more efficient way to produce accurate reports on each campground, the bookkeeper added the job-costing module to her BusinessWorks (State of the Art, 800-854-3415) accounting software.

Job costing, commonly used in the construction industry to track a job from the bidding stage through completion, lets contractors break out profits or losses on a job-by-job basis. But the term "job" can have a much wider application.

The bookkeeper realized that her campground reporting needs could be structured as a job-costing problem. By defining each campground as a job, she was able to track expenses for cleaning, maintenance, labor, and miscellaneous items against revenues for each campground. The system was easy to set up, and the results have been excellent.

GOING PUBLIC

Electronic High Flyer

Robert Kopstein's goal was to get as many prospectuses into the hands of as many potential buyers as possible. So he turned to technology to boost the initial public offering (IPO) of Optical Cable, a fiber-optic cable manufacturer in Roanoke, Va. Kopstein opened a Web site, posted his intentions, and booked tombstone ads in major newspapers—stark announcements that, by SEC edict, do no more than invite a reader to send for a descriptive prospectus.

He also distributed 10,000 prospectuses to his neighbors and assigned a cadre of employees to staff the phones, "like a bunch of telemarketers." Over four weeks, they booked orders for 670,000 shares from some 850 buyers.

Using the Internet and a phone bank to launch the IPO proved so successful that Kopstein wanted to sell another 500,000 shares. But the earlier buyers got restless and pushed for stock to trade on the open market. The stock of the $36-million company made a spectacular debut on Wall Street, reaching a high point of $136 a share during its first two months of trading.

SUPPLIERS

A Sip of the Action

Like many other small-business owners, Bill Murphy is finding that it costs little to establish a presence on the Web. The founder of Clos LaChance Wines, in Saratoga, Calif., opened a Web site to compete with larger wineries like Robert Mondavi. “We’re a little-bitty guy,” says Murphy. “We cannot compete from an advertising standpoint with bigger wineries. But on the Web,” he says, “our information will be as accessible as theirs.”

The company pays $50 a month to **rent space on a computer connected to the Internet**. Murphy’s Web storefront is drawn from a marketing brochure a friend made computer-ready at no charge. Where Murphy really saved money was in the deal he made with the developer. “If the page starts to bring in new customers, I’ll pay him a percentage of sales,” notes Murphy, who paid the developer’s up-front fee in wine.

Just because the Web has millions of users, it doesn’t mean that millions of people will suddenly start buying Murphy’s wine. The challenge for the designer of any small-business Web site is to turn surfers into customers, and that is why Murphy is smart to give his designer incentives to do just that. Buttons on Clos LaChance’s site, for instance, connect to pages called News, Overview, Awards, Catalog, Order Form, Guest Book, and What’s Popular. Murphy hopes that when visitors read, for example, that his 1992 chardonnay sold out after receiving favorable trade reviews, they will want to buy his wines.

SUPPLIERS

Ease Your Vendors into the Info Age

If you have to **link up clients and suppliers**, groupware may be your answer. When he designs a hospital or medical center, David Johnson, co-owner of Johnson Johnson Crabtree Architects, works with as many as 50 subcontractors. His Nashville-based business is healthy, generating $1.5 million in annual billings, but mailing diagrams, blueprints, and reports to clients and getting them back again can take days. And there are always complications, says Johnson. Consider "the problems and confusions when there are 15 people in 15 far-flung locations. Although they all get copies of important material, somehow they can't find the documents when you finally get hold of them."

For about $60,000, which bought design, installation, training, and a server, Johnson had a consultant shape Lotus Notes (Lotus Development, 800-343-5414) into a system that would fit his business. It took his staff of 12 approximately two months to learn and master the basics. The system incorporated every phase of the work process, linking the building owner, engineers, and the contractor's main and field offices. It works like a charm, ensuring that everyone sees copies of the same drawings and product samples.

But the key to success was signing up contractors and other suppliers. Johnson called a meeting of about 20 outside engineers and subcontractors. "We believe this is the way to do business, and we'd like you to join us," he told them. As an incentive, he offered to help install a free copy of Lotus Notes and to provide monthly training sessions, which cost him about $1,500 for each location. Ten people signed up right away.

SUPPLIERS

Put Your Phone Bill Up for Bid

Rykodisc, in Salem, Mass., is a midsize record company with offices scattered around the world. Its management relies on telecommunications to keep everyone connected. Of course, all those phone calls from one remote office to another can get expensive. But some smart buying decisions not only help keep costs under control, but actually reduce them.

In the early 1990s, at the suggestion of a telecommunications consultant, Rykodisc began **putting its long-distance service out to bid**. It's a tactic that most businesses use with raw goods suppliers, but few managers think to apply to services. But heavy competition in the telecommunications industry means that Rykodisc's gamble paid off. The company's cost per minute dropped from 16¢ to 13¢ in the first year, and since then, the cost has dropped even lower, saving the company more than $10,000 annually.

V

"With a telephone line,
you can compete
with anybody,
anywhere."

LEWIS FULLER
president of Fuller Medical,
Gadsden, Ala.

BACKUPS

A River Runs through IT

When the Red River overflowed its banks and flooded Grand Forks, N. Dak., it displaced more than 50,000 residents and devastated virtually every small business in the downtown area. But because Howard Palay had automated seven years earlier, at least one longtime Grand Forks business owner was able to **keep Mother Nature from crippling his company**.

Palay's 17-employee mail-order company, Palay Display Industries, sells retail fixtures and supplies. Customers place orders using toll-free numbers that ring in either Grand Forks or a branch office in Minneapolis, operators punch data into computerized customer files, and labels are automatically printed out in a main stockroom.

Once it became clear that Grand Forks would be evacuated, Palay knew what he had to do. He loaded his family and the server holding the company's records into his minivan and hightailed it to his Minneapolis office. That office uses only dumb terminals, which are connected to the main server in Grand Forks by a 56-Kbps direct line. For inventory, he would have to rely on the stash of fixtures in Minneapolis. "If the server had gone under, we would have been done," says Palay.

When Palay reached Minneapolis on Sunday, he spent about four hours rigging the dumb terminals to the server. With some help from his Minneapolis-based hardware supplier, Palay had rebuilt the network by 9 a.m. Monday, and the company was up and running as if nothing had happened.

Quick Backups for Big Data

At Cambridge Pragmatics, in Somerville, Mass., founder Sanford Friedman stumbled upon a fast and inexpensive data backup method when he bought a CD-ROM recorder to make product prototypes. The company started using the **CD recorder to back up company data**.

"The CD recorder is faster and easier for backing up and restoring files than a tape backup device," Friedman says. CD-ROMs take little more than half an hour to fill, minutes to read, and, by calling up a directory exactly as you do with a floppy or hard disk, it takes only seconds to find a file.

Friedman says a CD writer will work with almost any computer that has enough empty hard-disk space for about twice the amount of data you want to put on the CD-ROM. The extra space is needed to make an image file of the CD-ROM before actually writing to the disk. There are faster systems that can, however, write directly to the CD-ROM.

Systems are available for less than $500, and you can buy blank disks for as little as $3 each.

BACKUPS

When Lightning Strikes

When lightning hit her server, it fried the network cards in 7 out of 11 of Sheila Skolnick's computers. "It was devastating," says the owner of Elite Companies, an $18-million hotel-supplies company in Setauket, N.Y. "Our system was out for two days." She survived the episode owing to **good backup planning**. Here are her tips:

- *Always keep hard copy.* While the system was down, salespeople sorted through hard copies of invoices. Looking up paper records meant that customers who were used to receiving quotes in seconds were told, "I'll call you back in 10 minutes with that price." That was better than, "I'm sorry, we can't help you. Our system is down."
- *Make smart backups.* "When you have everything on the computer and you sell thousands of products, like we do, you get really dependent on your technology," Skolnick warns. She'd been backing up all her data weekly, sending the tape to a bank vault for safekeeping in the event of fire, flood, or theft. The data were ready to be reinstalled when the system came back up.
- *Buy the protection that's right for you.* Once the system was up and running, Skolnick was tempted to buy surge protectors that would cost about $100 for each computer, but she decided that was really more than she needed. After all, it's hard to protect against a direct lightning strike. Instead, she invested $25 for each computer's Newpoint surge protector (Newpoint, 800-639-7646). Each of her protectors came with a $10,000 insurance policy in case it should fail to stop lightning damage.

BACKUPS

Long-Term Plans for Active Archives

Everybody knows that for protection against disk failures, viruses, fire, and theft, you need a daily backup system. But what about information, like tax records, personnel data, customer lists, or historical profit-and-loss statements? You might want to **retrieve those records in 5, 10, or even 50 years**. You need to take into account not only the reliability of the medium but also the problem of obsolescence. Here are a few tips for long-term backup:

- *Use the right medium.* Digital tape is unreliable for long-term storage. It stretches, sticks to itself, and tangles. Erasable backup media—magnetic disks and rewritable optical disks—are risky because they could possibly be erased. As the price of CD recorders has dropped, it's become the long-term backup medium of choice. Manufacturers claim that CD-ROMs last from 70 to 100 years.
- *Think about how you'll access your data.* Your backup strategy should include both software and hardware so that you'll be able to retrieve the data. Whenever you upgrade software or an operating system, think about converting your long-term backup files. Most programs can read only the previous version or two, and you don't want your files to be beyond your grasp.
- *Don't forget hardware.* Optical drives that can read CD-ROMs should be widely available for at least 10 years, but a computer 50 years in the future is unlikely to use any storage medium with moving parts. So whenever you upgrade your hardware, also think about transferring your backup data to an upgraded medium. You don't want to be haunting flea markets in search of an antique computer that can open your files.

BACKUPS

Get Coverage for a Computer Crash

Linda Lewis admits that the worst technology mistake she ever made was not having **virus-scanning software and insurance against computer malfunctions**. The CEO of Plantworks, a $2-million interior plantscape-design company in Las Vegas, remembers when she realized her error:

"In 1992, while I was in South Africa visiting a client, the Michelangelo computer virus swept across the United States. My company was one of a few in Las Vegas that it hit. Our entire computer system crashed. Not only was I out of town, but I was unreachable because there were initially no phones at the customer site. My office manager didn't know where I kept the backup information and original program disks.

"Finally, about a week after the virus hit, I got to a phone and discovered what had happened. I'd heard of the virus because it had made the papers in South Africa. At first, I thought we'd lost only a couple of days of information. But we had nothing—nothing—left. The office manager had called our insurance company to see if it would cover the cost of getting the system back up and running. That kind of coverage is available, but we didn't have it.

"You don't want to know how much it cost us to fix things. I probably lost an additional $6,000 in man-hours reloading all the programs and rekeying the data. We spent hundreds more for the computer consultant we hired to help us and for new software, including a virus-scanning program. Now, I have insurance that covers computer malfunctions. I mean, come on. For $50 a year, it's worth it."

BACKUPS

Say Good-bye to File Cabinets

Adrian Stern and his partners had talked about switching to a paperless, electronic filing system someday. Then Mother Nature stepped in. An earthquake hit the Encino, Calif., office of Clumeck, Stern, Phillips & Schenkelberg, bringing chaos to the accounting firm's file room. With papers strewn across the floor, the partners decided the time had come to make the big transition: They would **store their old records on optical disk**. They decided to use WORM (write once, read many) disk technology, a relative of CD-ROM.

To switch over, the $2.5-million firm bought a medium-speed scanner, a dedicated computer with a WORM drive, an extra-large monitor, and a database program for records storage and retrieval.

With the change, the 14-person company no longer needs to rent extra storage space elsewhere in its building. That saves about $350 each month, and nobody can even remember how much time the partners wasted having to trek up and down several floors when they needed old records. Because the accounting firm already had an alphanumeric system for identifying old files, the business had only to transfer that system to the new storage mechanism. A part-time employee scanned the files and typed in their record numbers.

Stern gives the system high accolades, saying, "I can search the files and find things quickly."

BACKUPS

Your Computer Is Calling

When your company relies heavily on technology, keeping computers up and running is a top priority. Cambridge Pragmatics, in Somerville, Mass., uses cellular telephones to ensure that someone is always available to tend the network.

If Cambridge Pragmatics, a supplier of forecasting and modeling software for managed-care providers, has a power failure, the company's backup power supply in the computer room automatically takes over. At the same time, **the computer places a wireless phone call to the technical person's beeper**. He can then dial in using a cellular phone and try to save things. If that doesn't work, he can shut the computers down as gracefully as possible. With the cellular data-communications capability, he doesn't have to lose valuable time running to a wired telephone to dial in. And the computer can always reach him, even if the company's phone lines shut down.

LEGAL

Why Pay a Lawyer? Use the Web

On August 16, 1997, Greenwich Consulting Group (GCG) held its first board meeting—at Yankee Stadium. The two founding members of the strategic-planning and telecommunications consulting firm, in Greenwich, Conn., typed their minutes into a laptop while below them the Texas Rangers tromped the Yankees, 8 to 5. The event was a celebration of the company's technologically enhanced incorporation the week before.

Vice-president Christopher Simmons had anticipated spending a lot of time investigating Connecticut's incorporation requirements, researching and reserving a company name, and preparing and filing a certificate of incorporation. Instead, he turned to one of a growing number of **incorporating-service Web sites** that perform all those tasks online. "We didn't want to mess around, so I got right on the Web," says Simmons.

After visiting a few sites, Simmons chose Corporate Agents (www.corporate.com). It took him 10 minutes to complete the application and enter his credit-card number. Corporate Agents did the rest. About a week later, Simmons had his company's official incorporation certificate in hand.

All told, it cost GCG about $550, $300 of which went to the office of the secretary of state. Had Simmons used a lawyer, he estimates that incorporating in Connecticut might have cost him as much $1,000.

LEGAL

Let Your PC Draw Up the Contract

When Munchkin Inc., a manufacturer and designer of infant products in Van Nuys, Calif., needed to prepare its second international-distribution agreement, its chief financial officer wanted neither to pay a lawyer to draw up a nearly identical contract nor to go it alone using the first contract as a template.

With lawyers costing $200 an hour, the do-it-yourself approach was tempting. Many managers resent paying lawyer fees for what seems to be a reinvention of the same old wheel, but doing it on one's own can lead to legal troubles that might cost more than hiring a lawyer in the first place.

Luckily for Munchkin, its law firm offered an appealing compromise: **interactive software that can automatically draw up distribution agreements**. The software prompts users through a series of questions about potential distribution agreements and uses responses to generate customized international-distribution contracts.

The program saves the companies from making costly gaffes. Consider, for example, the question regarding the length of contracts. The program asks how long the agreement is to last. If the time period the user enters exceeds the target country's permissible limit, the software will flag the discrepancy. The application also cautions users about clauses that might put either partner at a disadvantage.

Munchkin had paid its lawyers between $3,000 and $5,000 to draw up a typical agreement. The software cost the company about $1,500.

LEGAL

10 Steps to Developer Contracts

The best insurance against a problematic computerization project is a **strong, written contract**. Lawyers and other experts in the field recommend that it include at least these 10 provisions:

1. The players, their responsibilities, and due dates. Be sure to designate who the decision makers are.
2. A detailed description of the scope and nature of the work, with deadlines.
3. A provision for progress reports or development updates. For payment purposes, break the project into phases with milestones that enable you to monitor quality before everything is set in code.
4. A compensation schedule, specifying hourly rates or flat fees, along with terms and conditions of payment. Reserve the final payment until after the job is finished and everything is working to your satisfaction.
5. A confidentiality provision making it clear that proprietary information (for example, customer lists and financial records) will remain confidential.
6. Copyright language.
7. A provision for free upgrades if the project involves software that is later improved or modified.
8. A warranty, stating that the work will be free from defects in workmanship and materials and conform to the project specifications. A one-year warranty is reasonable, but some vendors try to limit their warranties to 90 days.
9. A flat prohibition against secret code, usually a "time bomb" or time lock, that could disable your system if payment is not made to the developer.
10. A provision for operators' manuals and training, if needed, before final payment is made. Also, specify the cost of follow-on training and service.

LEGAL

Intranet Solves Record-Keeping Problem

How does one small organization use an intranet, or internal Web site, to keep human resources operations up to par? Before Forsyth Dental Center, a nonprofit organization in Boston, launched an intranet, Forsyth's 150 employees had been working fairly independently. "There wasn't much centralized management, and the **lack of good record keeping ran us into some legal problems**," says Douglas Hanson, the center's director of computing and network technology.

To improve the organization of such data as research guidelines, grant-application updates, and employee policies, Hanson had considered the popular Lotus Notes groupware (Lotus Development, 800-343-5414) for sharing data files. But he soon realized that Notes was way too expensive for Forsyth's 40 computer users.

Instead, Hanson decided to install an intranet. At the time, it cost $400 to buy the additional Microsoft Windows NT licenses that would allow many employees to access the site simultaneously. "We have only two people running the network and not a lot of money," says Hanson. He chose FrontPage (Microsoft, 800-426-9400) for creating and updating the site, and he encourages department heads to contribute to the site and keep their sections up-to-date.

LEGAL

Is E-mail Private? Not at Work

Is employees' e-mail protected by privacy laws? Recent controversial court rulings have generated confusion on the matter. Lee Gesmer, partner in the Boston law firm of Lucash, Gesmer & Updegrove, answers key questions about the legal aspects of **monitoring employee e-mail**:

- *What rights to privacy do employees have on e-mail?*
  The law provides little or no guarantee of personal privacy on a company's e-mail system. There is no constitutional "right of privacy" for internal e-mail communications. The Electronic Communications Privacy Act of 1986 prohibits the interception of messages sent over online systems or the Internet, but exceptions exist that give employers the right to monitor employee e-mail.
- *Is it legal to read employees' e-mail without their knowledge?*
  Yes, under most circumstances. The issue has arisen in cases in which an employee has been terminated after the employer read e-mail. To date, no court has upheld employee arguments that termination was based on a violation of privacy rights.
- *What is an effective e-mail policy?*
  Avoid giving employees the impression that they can expect e-mail privacy. This message should be stated in policy manuals and posted prominently on the e-mail system itself. However, the law is so favorable to employers that even when they do the opposite—tell employees that e-mail will be treated as confidential—they may not be bound by their promise.
- *How should companies handle e-mail monitoring?*
  That's up to the employer and based on technical and personnel resources. There are no legal restrictions.

LEGAL

Who Owns Your Web Site?

If you're not careful, someone else could end up **owning your Web site**. Be sure to address these issues early on:

- *Who has the intellectual-property rights?*
  A hired designer may want to retain the copyright to elements such as icons, graphics, or layout. Expect to pay more for designers who are willing to part with copyright. In addition, if you aim to include scanned-in artwork from any outside source, you should negotiate the use of material protected by copyright. And, if your site too closely mirrors another, you could be looking at trouble. Recently, White Rabbit Toys, a small toy store chain based in Ann Arbor, Mich., sued another company for copying the look of its site.
- *Who owns the code?*
  If you use the proprietary software of your Internet service provider (ISP) to design your site, the ISP may not let you take certain elements with you should you leave. That's because those Internet specialists own the "architecture," or software that supports what you've designed. You can negotiate for a license to use the software permanently or temporarily once you leave.
- *Who gets the data?*
  The data you collect at your site might not be yours to keep unless you have an agreement specifying ownership. Your contract should also forbid your ISP from showing your data (for example, hit rates or transactions) to anyone but you. An ISP may be tempted to use such data to impress prospective customers.

LEGAL

Software Improves Managerial Assessments

Kerry and Kevin Schulz of Vidcon Enterprises, in Battle Ground, Wash., say that performance-evaluation software did more than **help managers write employee reviews**; it made them better managers. The program was teaching managers at the book-and-video convenience-store chain a more effective language for evaluating employees—language that is objective, to the point, and consistent. Whenever managers slip and write something inappropriate, the program alerts them to the mistake, making them aware of important policy distinctions.

If, for example, a manager writes, "This employee is too young for the position," a box appears, warning against confusing experience with age. When managers are puzzled about certain terms, they can click on an advice section for additional information.

The software Vidcon relies on is Performance Now! (KnowledgePoint, 800-727-1133). It features a series of 30 "elements"—categories like job quality and overall cleanliness—for rating employees.

LEGAL

Automate HR Problem Solving

Need help tracking employees' performance? Confused about how to comply with government regulations? Wondering whether it's appropriate to ask an applicant how she spends her spare time? Human resources software can fill in for well-trained HR professionals, helping small companies **troubleshoot legally tricky HR issues**.

William Floyd, executive vice-president of Investors Financial Group, a $70-million financial services provider in Atlanta, reviewed HR Task Counselor (Jam.LOGIC Designs, 800-750-8113). "A few years ago, I would have paid dearly for this software," he says. "Since then, I've hired a full-time HR person. She's looked at several packages that cost two to five times more than this one and don't provide as much information."

If you employ from 15 to 200 people, and human resources issues are overseen by the controller, the office manager, or even you, HR software might be a real boon. It enumerates questions that you may and may not legitimately ask job applicants, helps managers prepare employees for evaluations, and offers a model exit interview, release agreement, and description of COBRA benefits.

"Not only do employers benefit from the computer's comprehensive reach, but the software lets employees check out new-job and other postings, benefits, and a company handbook we can update at will." Floyd says the software makes life easier for his HR person and enhances the employees' attitudes about the company.

LEGAL

Respect the Net's Power

The Internet might be the cornerstone of your company's marketing strategy, or it might facilitate communications among employees spread across the country. It also might get you sued. According to Barry D. Weiss, a partner at the Chicago law firm of Gordon & Glickson, if you **give employees access to the Internet, give them specific guidelines, too**. Traditional corporate communication policy, says Weiss, can't account for a medium in which a libelous letter can reach a million people in a matter of minutes. He advises companies to take the following precautions:

- *Implement authorization codes.* Identify employees who are allowed to use the Internet, and give them passwords to get online.
- *Control participation on bulletin boards.* To reduce the risk of defamation claims, restrict participation in chat channels and bulletin boards to employees who have specific business to conduct in those forums.
- *Put your rules in writing.* Employees should sign agreements stating that they understand that unauthorized use of the Internet could be grounds for dismissal.
- *Give advanced notice of monitoring.* To avoid being cited for invasion of privacy, be sure to inform employees clearly and in advance if you plan to monitor their e-mail or Internet use.
- *Limit downloading of information.* To protect against copyright-infringement liability, allow employees to download or distribute only those Internet materials that include a copyright notice and specifically permit such dissemination.
- *Pass by legal.* Have your lawyer review your formal Internet policy.

OFFICE

Choose the Right Network for Your Company

Choosing between a peer-to-peer and server-based network operating system for your local area network (LAN) is akin to deciding whether to go Macintosh or PC on your desktop. Peer-to-peer networks are created by stringing computers together in a loop; server-based networks connect computers to one central computer, the server. Although each has distinct advantages—and loyal users willing to wage war for their favorite—the two systems increasingly resemble each other. Unless you are a technical wizard, the choice is best left to your networking vendor. That said, consider the following pros and cons:

Peer-to-Peer Systems

Pros: Less expensive
Easier to install and maintain

Cons: Performance declines with more than 25 users
Difficult to manage with more than 25 users
May be less secure
Individually administered

Server-Based Systems

Pros: Expandable to support hundreds of users
Centrally administered
Powerful security

Cons: Complicated to install and maintain
More expensive

OFFICE

Digitize Your Documents

Imaging technology—the combination of software and hardware that turns paper information into digital data—has earned a pretty bad rap for being expensive, unreliable, and difficult to operate. Lately, however, the products have improved, and prices have dropped. Imaging technology is now a reasonable way to **store documents, conserve space, and save money**.

"Until a few years ago, imaging technology was a farce. But now it's opened a whole new world," says Roger Gamblin, president of Flagler Title, a title-insurance agency in West Palm Beach, Fla. For years, as part of each preclosing review, Gamblin's staff manually assembled and collated hefty packets of legal documents and insurance papers for shipment to the firm's clients. An attempt, in 1988, to replace that cumbersome process with imaging technology failed: available products were slow and suited only to archival storage.

These days, however, when it comes time for preclosing reviews, Gamblin's company uses PaperPort Deluxe software (Visioneer, 800-787-7007) and a slew of scanners to process the necessary documents. An employee records the digitized material onto a CD-ROM and sends each customer a single disk in one slim envelope. Three to eight parties are involved in a typical deal, so the cost of labor for scanning in the hard copy and making a disk is more than offset by the $500 or so each of those clients saves in copying charges and overnight-mail fees.

• OPERATIONS •

OFFICE

File It and Find It

To save himself from drowning in paper memos, reports, and documents, William Floyd, executive vice-president of Investors Financial Group, in Atlanta, did away with his paper filing system altogether. The $70-million company relies on an **electronic filing system** to store every document, and every employee has access to the system and its powerful search features.

Floyd first scans every document into the computer. In the next step, filing the documents by category, the images are converted into graphic files by PaperPort (Visioneer, 800-787-7007). The software automatically extracts keywords from the document and can add any others that Floyd types in himself. When they want to retrieve documents, Floyd and other employees can search by title, content, or keywords.

The software automatically files documents, searching the database for folders containing documents of a similar category—policies, receipts, correspondence—or with similar layouts. Based on its search results, it suggests where the new document should be filed.

"It's a wonderful tool for finding documents while I'm traveling," says Floyd. "The file folders save me considerable connect time when I'm looking for something I filed six months ago to help close my next deal."

OFFICE

Peanuts, Popcorn, or Power Tool?

When you stand in front of a vending machine pondering your choices, you're probably thinking "Snickers or M&Ms?" not "power drill or earplugs?" But at Prince Industries, in Carol Stream, Ill., employees benefit from a vending machine that dispenses tools.

The Automatic Tool Dispensers (ATDs) are **computerized vending machines** that allow workers who punch in the correct codes to take and return tools they use on the shop floor. The same system could also work with laptops or pagers. ATDs can be as small as a bread box or as large as a refrigerator, and their prices can range from $15,000 to $50,000 (Vertex Technologies, 800-249-4933). The machine's software tracks the tools and their borrowers, and because they can be placed right on the shop floor, the ATDs are easily linked to desktop computers. It's easy for managers to run reports on inventory and usage.

Bob Trebe, plant manager at Prince Industries, a $12-million contract manufacturer, uses the vending machine's reporting capabilities to keep an eye on how workers are using tools. When a daily-usage report revealed that one worker was ripping through turning-tool inserts during a certain job, Trebe decided to investigate. He discovered that the employee had the machine running so fast that he was wearing out the inserts, so Trebe demonstrated how to slow the machine and extend the life span of the tools. He estimates that the ATD saves his company thousands of dollars a year in inventory management. And it cuts the time workers would otherwise spend walking to and from a tool crib—up to four miles a day, according to one study.

OFFICE

Will the Year 2000 Bring a Crash?

The year 2000 could cause serious problems for software that uses two-digit date fields. A date entered as "03-01-00," for example, can crash software packages that recognize "00" as 1900 rather than 2000.

That glitch might affect companies that make heavy use of financial software, manufacturers that use electronic data interchange (EDI) software, or even companies projecting sales over the turn of the century.

Find out whether or not you have a Year 2000 problem. If you use custom software, you should contact the programmers and ask whether their products are "year 2000 compatible." If not, find out whether they have developed a solution, and try to schedule a repair. "Smaller companies running off-the-shelf software can usually replace just those modules," says Lynn Edelson, a partner specializing in computer issues at Coopers & Lybrand, in Los Angeles.

To check operating systems for errors, reprogram your computer's date and time to read "12-31-99, 11:58 p.m." Then turn off the computer and wait three minutes. When you turn it on again, check the date. If it does not say "01-01-00," you've got a problem.

OFFICE

Online House Calls Save the Day

It's a busy afternoon, and the installation of a software upgrade isn't going well for Grafton Associates, in Kansas City, Mo. Richard Carroll, CEO of the $10-million temporary-help agency, calls in his computer specialist, who reinstalls the software while the staff watches. A few minutes later, the program is working fine. It's a normal scene in corporate America—except that Grafton's computer specialist is 1,198 miles away, in New York City.

Carroll chose the **outsourcing solution**—he pays $75 an hour for a few hours' help each month—because he didn't need a full-time MIS manager. But the arrangement didn't start efficiently. When he had to telephone or e-mail the consultant describing the problems, Carroll explains, "He had to imagine what was going on."

Finally, Carroll invested in a dedicated modem line and modems for his two locations. Now the consultant dials Grafton's modem, logs into the company's Apple network, and watches while a Grafton employee reenacts the problem. If the problem isn't urgent, the consultant waits until the workday ends and modems in to fix it. When the staff arrives in the morning, the problem has been solved. Says Carroll, "We're free to do our core business—finding temporary employees, not fixing glitches."

• OPERATIONS •

OFFICE

Smarter Network Setup Stops Surges

Sheila Skolnick thought she had plenty of protection. Surge protection, that is. The owner of Elite Companies, an $18-million hotel-supplies company in Setauket, N.Y., had been using computers to track sales and inventory for years, and, she says, "We thought we had taken all the precautions." The company used a standard-strength uninterrupted power supply and surge-protection strips.

But none of that saved her from the lightning bolt that zapped her company's computers at 3 a.m. one Sunday. The lightning wiped out the network cards in 7 of her 11 computers and taught her to **set up her network differently**.

Skolnick's network had been wired together in series, with the server at one end. As a result—like a string of Christmas tree lights when one is burned out—even the unharmed computers wouldn't work. Now her server is flanked by six computers on one side and five on the other. That way, if lightning strikes twice, it will have a shorter distance to travel before going to ground. At worst, it will wipe out only six computers.

OFFICE

ISDN Lines Link LANs

When the Main Street Cos., a 125-employee restaurant and catering company in Princeton, N.J., wanted to link the local area networks (LANs) of its three sites, vice-president John Marshall considered several options for the $5-million business. Marshall, who had read about ISDN technology in a computer magazine, contacted his local phone company about the service.

Bell Atlantic provides him with **ISDN lines that link the company's two restaurant-coffeehouses to its production facility**, and it costs much less than the other available wide-area networking options. Marshall estimates that the next-cheapest alternative would have cost Main Street $800 to $1,000 a month in addition to initial hardware outlays of some $15,000. In addition to the $3,600 he invested in hardware, Marshall pays $80 a month for his ISDN service.

Marshall says ISDN was easy to install, involving only two big changes. For each building, he needed an NT1 (a device that hooks the company into ISDN), and each server needed an ISDN card, which, he explains, "you plop in like a modem." Now all three sites share databases, information, and staff expertise.

• OPERATIONS •

OFFICE

Host Your Own Web Site

Before you start to design your Web site, you should decide whether you will store it in-house or with an Internet service provider (ISP). **Serving a Web site in-house is more expensive, but it can be the smarter choice in the long run**.

When the computer that hosts your Web site is just down the hall, it's cheaper and faster to experiment. You can change, update, and expand your site at will. If you want to gather customer feedback, update price listings, and tweak the site as your company grows, you can do so quickly, easily, and economically with an in-house server. You can also link the site directly to such company information resources as sales, inventory, or customer-service applications. If a visitor to your site should suddenly decide to order 10 million widgets, the on-site server can link the sale directly to your order-processing software.

OFFICE

Bringing Your Site Back Home

Here are some tips from Phaedra Hise, author of *Growing Your Business Online: Small-Business Strategies for Working the World Wide Web* (Henry Holt, 800-488-5233, $14.95), **for a Web site that's going to be stored in-house**.

- A dedicated phone line is a basic component of your Internet service. Because phone company orders are notoriously backlogged, you may have to wait a while to have your phone line installed. Contact your Internet service provider or local telephone company as soon as you begin considering the setup.
- Specialized "Internet server" computers are overpriced and over-powered for most sites. For an entry-level site, start out with a smaller, less expensive Pentium-based computer with a fast modem. If customer demand overpowers that machine, you'll be able to justify the expense of something bigger.
- Determine who will set up and maintain the system. Will it be an employee, an outside contractor, or some combination of the two? Someone needs to be on call at all times; if the system crashes, you want it back online as quickly as possible.

OFFICE

Getting a Fix on Outside Contractors

Air Taser, a stun gun manufacturer in Scottsdale, Ariz., decided the best way to handle order fulfillment was to outsource. If you outsource any process, you'll find that **good contracts make good partners**. Here are five issues critical to any outsourcing contract that Air Taser founders Rick and Tom Smith addressed in their contract:

- *Pricing and compensation.* Spell out how services will be billed. Will the customer pay by the hour? By unit of computing power? By number of items handled? Vendors should make provisions for changes in their own costs over the long term. Include payment schedules based on clearly defined criteria.
- *Description of services.* Define every service in detail, and stipulate which side eats costs in case of problems. Vendors should specify tasks that are part of the core service and those that are extras.
- *Performance standards.* Describe what metrics—transaction times, reporting cycles, and so forth—both parties will use to measure performance. Build in flexibility. These are long-term relationships, and the world is always changing.
- *Management contact.* Small companies can't have every manager supervising outsourced relationships. So set up a mechanism or designate a special liaison who will keep the partnership on an even keel.
- *Escape routes.* Plan for the end. When the relationship is over, what happens to inventory, equipment, and mailing lists? What must the vendor hand over to the new outsourcer or to the customer? How much notice must each party give before terminating service?

What You Should Know about ISDN

ISDN telephone lines are supposed to be the next major front in the digital revolution. An ISDN line moves information at the fast speeds and in the digital language that computers prefer. When things work well, **ISDN lines can transmit at rates as fast as 128 Kbps**, roughly nine times faster than that of a 14.4-Kbps modem. But there are still several key problems with ISDN lines.

Whether you can have one installed depends on whether the software that controls your local Baby Bell's switches meets national ISDN standards. Then, of course, there's the question of access. In the analog world, if you want to have a good chance of getting online whenever you want, you should aim for an Internet service provider with no more than 10 customers for each of its modems. In the ISDN world, the key ratio is B channels. Each B channel represents an available connect speed of up to 64 Kbps. You have to connect on two B channels to get full speed, which isn't an easy task in the face of growing customer demand.

In addition to a $350 digital modem, you'll need a high-speed serial port. ISDN calls are almost always toll calls, even if they're local. Some phone companies charge as much as $200 to install an ISDN line, but hourly connect charges are roughly comparable to analog rates. Monthly ISDN Internet costs often come to less than conventional analog dial-up costs because you can get more done in less time at ISDN speeds.

Still, with high-speed cable modems moving onto the market horizon, there's the inevitable question of when ISDN will become obsolete.

OFFICE

Wire Your Community

If your industry is slow to come online, maybe you can help it along. You can bring businesses together using e-mail or the Internet. And, because you'll be using fewer faxes and overnight mailings, staying in touch will cost you less. Before you leap into cyberspace, you can lay some groundwork **for bringing your industry online**:

- *Organize.* Consider a new organization to provide Internet access and training to your target companies.
- *Do outreach.* Talk to people and community organizers face-to-face. Explain how the Internet can help their businesses.
- *Educate.* Provide quick-start Internet classes to develop a critical mass of techno-literate employees. Some students may end up helping out at the online service provider or training other students.
- *Target.* Offer workshops for specific business groups to demonstrate the specific benefits of going online.
- *Give it away.* Free public access to the Internet at a local library or community center lets people get their feet wet gradually. Proprietary online services like America Online give away computer disks with connection software and offer free service for the first hours of use.

OFFICE

Play Computer Games to Sharpen Skills

Sneaking in a daily round or two of Doom might actually help you run your business better, says Lewis Paine of Opta Food Ingredients, in Bedford, Mass. The CEO of the $10-million company, which makes natural food ingredients, regularly indulges in sessions of SimCity (Maxis, 800-33-MAXIS), a computer-game classic.

SimCity challenges players to make executive decisions that affect an entire community. "For example," Paine explains, "if you build a housing development, you're going to have to build roads, and that comes out of the budget. SimCity allows you to see a multitude of factors, such as city planning and budgeting, not only quantitatively but visually."

How does **playing computer games** help him run a company? "It's a phenomenal teacher of how to allocate resources correctly and anticipate the needs that arise by taking one particular course of action," says Paine. "After you play a few times, you realize that you have to get better and better at anticipating the future impact of whatever you're doing now and really think through the allocations of the limited resources you have."

"The most effective executives have developed personal productivity regimens that exploit technology while clearly setting limits on how much they let it control their lives."

JOE PHELPS
CEO, Phelps Group,
Santa Monica, Calif.

SECURITY

Lock-less Monster

Now that everything is digitized and access is password controlled, employees can waste an awful lot of time creating, remembering, and changing passwords. A company manager who wishes to remain anonymous considers password management:

"For security reasons, everything has a unique password. They don't actually come out and say, 'You have to forget all your passwords so you can't do any work.' What they say is, 'You have to keep changing your password every 10 minutes for security reasons.' And, 'Make sure you pick a password that isn't anything like your other passwords, or a common number, or an ordinary English word.' And the perfect suggestion to **make you forget all your passwords**, 'Make sure they aren't anything anyone else can guess.'

"If you follow all these rules, no outsider can get into your voice-mail box, computer, fax and copy machines, parking garage, or bathroom. And neither can you.

"You have to spend your day thinking up new, alien, and unmemorable passwords and then trying to reconstruct how you invented them. 'Well, it was something I'd never said or heard, like *smergreb*. But it wasn't that because I've never heard or said that before, not even 20 minutes ago.' So here we have a national problem. At my company I am paid not to write a staffing plan, not to create a sales brochure, not to evaluate suppliers, or to interview customers. No, no. I am paid to think up words like *spuglucz*, that have no memorable qualities. And when I forget *spuglucz*, then I am paid some more to stare at the screen and invent words even more valueless."

SECURITY

Update Your Virus-Protection Software

For years now, Alex Haddox, of the Symantec Antivirus Research Center, in Santa Monica, Calif., has studied the devastating effect of boot-sector viruses. Those viruses infect the files that allow the computer to operate, and they can be transferred only by human contact, that is, by floppy disk. But boot-sector viruses don't keep Haddox up at night. There's a new kind of virus that's even worse.

The latest threat, dubbed a macro virus, burrows into the code of such applications as word processors. Unlike its predecessors, it can easily migrate across most platforms. Their devilish lines of code, simple to write, can sit dormant on a computer for years before they bring down an entire network. Worst of all, a macro virus could spread across the Internet at a terrifying pace.

The first macro virus to be identified was called the "concept virus." It's the most prolific virus in history. That particular virus does no harm beyond printing a secret message, "And that is enough to prove my point," in the computer's code.

Macro viruses get mainlined into computers around the world every day. A bandit infected a computer at the University of Auckland, in New Zealand. The Antivirus Research Center was summoned, but in only two weeks the virus had spread to six countries. "We cured it, but not before it had reached as far as South Africa, England, and the United States," says Haddox. It's a good argument for using and frequently updating your company's virus-protection software.

Protect Your Computers: It's Worth the Price

Computer security isn't always expensive. Break-ins usually are. Simson Garfinkel, author of *Practical UNIX and Internet Security* (O'Reilly, 800-998-9938, $39.95), offers the following advice for small-business owners who want **simple security on the cheap**:

- *Develop a written computer policy for your company.* State whether or not employees are allowed to log into the company's system from home. Also, clearly state who is responsible for backing up computer files and whether the company will monitor employees' e-mail.
- *Minimize exposure to the Internet.* If you want your company to have access to the Internet but don't want outsiders to have access to your internal information, try "air gapping." Put simply, air gapping means connecting only one computer to the Internet and making sure it's not connected to your internal network. That way, if someone breaks into the connected computer, there's no way to reach your internal system.
- *Back up everything.* No security measure can guarantee 100% protection from a break-in or such natural disasters as fire and water damage.
- *Set up internal firewalls.* A firewall is software that limits who is allowed into and out of your system. A lot of companies erect only external firewalls because they fail to understand that most computer crimes are inside jobs. Internal firewalls can protect sensitive information from the wrong insiders.

• OPERATIONS •

SECURITY

Scan Every File

Robert Slade, author and computer virus expert, explains **what viruses are and how to beat them**:

A virus is any program that reproduces itself by using the resources of your computer without your knowledge or consent. Most are not intentionally malicious. A truly malicious virus—say, one that erases an entire hard drive—doesn't have much chance to spread because it destroys its "host." But any virus you get will eventually conflict with something on your system. People think viruses have something to do with pirated software and that if they don't have a modem, they're safe. And everyone thinks, "it can't happen to me." None of those three things is true.

First of all, back up regularly. Second, everybody should have antiviral software; there's no excuse not to get it. I think shareware is often better than the commercial products. Many antiviral software vendors have introduced Windows 95, LAN, and Internet versions of their software, but they cannot guarantee protection. You should check each file attachment or Web-based installation you receive. Check every disk and new program you receive. Some antiviral software can monitor your system automatically, either all the time or at regular intervals. And finally, keep your antiviral software up-to-date.

You can remove most viruses by running good antiviral scanning software. If you can't fix the problem easily, confirm the virus with a second brand of software—it is possible to get a "false positive." If necessary, delete the infected file and reinstall it from your backup. Then scan all disks with the antiviral software, or you may reinfect your computer.

Firing Up Double Protection

There are **two types of protection**: firewalls and encryption. A firewall controls who gets into and out of your network. An encryption program prevents unauthorized people from reading your e-mail or the files on your system. One or the other—or both—may be right for your company.

Firewalls act as gatekeepers between a company's internal network and the outside world. At minimum, firewalls examine the location from which data enter your system or the location to which data are going. Then, based on your instructions, they choose whether to allow the transfer of that information. For example, you might set up a firewall to accept files from your office in Hawaii but to reject any other files. The most thorough firewalls also examine the contents of files for viruses, monitor the use of the system, and keep logs so that you'll know if anyone tries to break in.

Secure as firewalls are, they can't repel intrusion 100% of the time. So if you're looking for the ultimate protection for your company secrets, consider encryption as a second line of defense. Think back to the breakfast cereal you ate as a kid. Sometimes there was a paragraph of gobbledygook on the back of the box that you could read only by covering it with a special piece of transparent colored plastic. That's essentially how encrypting a computer file works. One person creates a message and turns it into gibberish, using a special "key," or code. Only someone with the right decoding phrase (equivalent to the colored plastic) can read the message.

• OPERATIONS •

SECURITY

Don't Scrimp on Hacker Protection

Jennifer Lawton, cofounder and CEO of Net Daemons Associates, in Woburn, Mass., received a disturbing phone call from an Internet service provider. The man accused Lawton of breaking into several Internet sites. Flabbergasted, Lawton assured the man that he was wrong. She couldn't believe that any of the employees at Net Daemons, which provides computer-network and system administration support as well as Internet services, could be hacking into other people's sites. But when Lawton and her cofounder, Christopher Caldwell, took a look, their worst fears were confirmed. Someone had broken into Net Daemons' system and, from there, launched attacks on several other computer systems, including those of some of its clients.

For the next six months, Caldwell worked with the Secret Service to put the ring of eight hackers into jail. The company lost thousands of dollars in labor: 10 hours a week working with the Secret Service and even more time restoring lost client data. Considering those costs, Lawton now believes that investing in a firewall is well worth the price, even for the $4.8-million company. Previously, Net Daemons had **carelessly relied on free firewall software** downloaded from the Internet. Chastened, the company spent about $5,000 to buy FireWall-1, a commercial firewall (Check Point Software Technologies, 800-429-4391). Until they were apprehended, the hackers continued trying to break into Net Daemons' system, but the new firewall stopped them every time.

"When I talk to clients about security, I talk about downtime," Lawton says. "An hour of downtime can cost thousands of dollars."

SECURITY

There Is No Natural Immunity

"Your data was delicious." That's the computer message that greeted Anthony Ferdaise when he returned from lunch one day. To his horror, Ferdaise discovered that every file on the hard drive of his PC had been erased, apparent casualties of a virus.

Ferdaise, who used to publish a fitness magazine in San Diego, turned to data-recovery software in a vain attempt to retrieve some of the lost files. Unfortunately, it had been more than two months since he'd backed up his hard disk. Ferdaise estimates that reinstalling all the software, re-creating his data, and scanning in all the photos he had lost postponed his magazine's launch by some six months.

He acknowledges having learned two valuable—if painful—lessons from the experience: **check regularly for viruses and back up religiously**. To those who think viruses won't strike them, Ferdaise cautions, "I said the same thing, and it happened to me."

• OPERÁTIONS •

SECURITY

Are You Suffering Password Overkill?

Allegra Print & Imaging, formerly American Speedy Printing, in Santa Clara, Calif., has six e-mail accounts. One is for general inquiries to the $1.6-million printing company, and five are assigned to people in the company. Each account has a different user name and password. Employees of the 11-person company are sometimes required to be out of the office and can't check e-mail, so inquiries were going unanswered.

For a while, vice-president Jim LeGoullon tried to check all accounts himself. But logging in and out of each account several times daily was tedious. His company had a case of **too much e-mail security**. He needed a more efficient way to manage the log-in process.

LeGoullon turned to Eudora Pro 3.0 for Windows (Qualcomm, 800-2-Eudora). It allows LeGoullon to set up one computer to check all the accounts every 10 minutes or according to any time frame he chooses. The program's Remember Password setting allows him to bypass all or some of the six passwords. That wouldn't work for all companies, he acknowledges, but at Allegra, everyone has access to all information, strangers don't have access to the terminals, and e-mail is used only for business.

SECURITY

Guarding Your E-mail

Paul Lewis, president of MC² Corp., a $10-million computer-network design company in Warren, N.J., lost customers and employees due to a **security breach in his e-mail system**.

"We discovered that someone was running rampant through our e-mail system. He had been reading all corporate e-mail and forwarding confidential e-mail between me and our vice-president to other employees, sending quotes and proposals to a competitor, and passing along payroll information to buddies. He once made up a message under my name that slammed the technical abilities of one of our engineers, and he forwarded it by 'user unknown' to the engineer.

"One night we broke the intruder's phone connection to our mail router and quickly traced the call. We tracked his movements for a few weeks to collect evidence. Ultimately, the police charged a former employee who had helped rebuild our e-mail system when it crashed one evening. He was ultimately convicted of two felony counts.

"Overall, we estimate the damages from that breach to be in the millions of dollars. We lost loyal employees when messages I sent to the vice-president about them landed in their e-mail boxes. We lost customers when pricing information was forwarded to a competitor who used it against us.

"Now we perform regular security audits, and we invested in three firewalls. We also have new policies about what information is allowed to be sent via e-mail, and we're extra careful about how we phrase our messages. We know that there's always potential for unintended eyes to see what we write."

SECURITY

Internet Dos and Don'ts

Companies are finding out that the Internet is a remarkable tool for business. But unlimited access by users of all stripes raises some serious issues. If you're looking to **play it absolutely safe on the Net**...

- Don't go overboard with your investment in the Internet. Expect any revenues you generate to be a very small part of your business for the foreseeable future.
- Be prepared to deal with hordes of users, who can clog up your phone lines or crash your server if your equipment isn't able to handle the traffic. If you want to limit access to authorized users, require your customers to use passwords.
- Assume that anything you send or expect to receive—including e-mail, credit-card numbers, and technical help bulletins—could be intercepted.
- If you plan to ask for credit cards, you should offer secure-server capability from a reliable vendor.
- Don't connect your internal Web site (intranet) to the Internet, even if both require passwords to access. This "air gap" is the only guaranteed protection from unauthorized Internet access.

SECURITY

Congratulations! We Found Your Weak Spot

Net Daemons Associates, in Woburn, Mass., provides computer-network and system administration support as well as Internet services. When its managers realized that someone had hacked into the company's computer system, founders Christopher Caldwell and Jennifer Lawton went into red-alert mode. Not only had the company's data been compromised, but so had data belonging to some of its biggest clients.

Lawton and Caldwell personally called the clients whose sites had been hit and gingerly broke the news. Their reactions surprised Lawton. "No one freaked out," she recalls. "Some were even relieved that the break-in happened." The attack alerted them to gaps in their security so that they were able to ward off future hits. In essence, the attack had served as a security check.

You don't have to wait for a similar attack to find the holes in your system's security. Hire a consultant, intern, or a techno-savvy friend to **try to hack into your system and discover its weaknesses**. You should discover problems sooner rather than later, and it's better to get the news from someone you trust.

• OPERATIONS •

SECURITY

Why Hackers Love Small-Business Networks

Most Internet users know by now that hackers can target any account. But the conventional wisdom is that hackers are only looking for "interesting" targets: phone companies, government agencies, large corporations with sensitive files, credit-reporting firms, or organizations with extensive credit-card files. That simply is not true. Many **hackers look for small, relatively unknown companies** with little of interest in their computer files.

Why? One reason is that small companies offer a perfect place for a new hacker to test his or her skills. Because defenses and security awareness are likely to be low at such sites, hackers have a far better chance of breaking in and rummaging around, and there isn't much chance of interference or risk of being caught. If your site is unintentionally accommodating to low-skilled hackers, you might win the honor of being labeled a "penetration test site" on hacker bulletin boards. Then you'll enjoy frequent visits from hackers all over the world.

For the same reasons, hackers also use smaller, less sophisticated systems as launching pads into larger, more security-conscious target sites. They mask their identities by first hacking into several smaller systems to create an electronic trail that's difficult to trace. Once they have what they want from the target site, they may then bring down all the sites in their trail to eliminate any traces of their presence.

"Word gets around fast," explains one hacker. That particular hacker has a day job—in the technical services department of one of the major online services.

SECURITY

Laptop Tips from the L.A.P.D.

Some estimates put the number of laptops stolen each year at more than 200,000, and that number doesn't include thefts that go unreported. In fact, businesspeople are easy targets because crooks know they usually don't report laptop thefts. But you can **keep your laptop safe** by following some basic guidelines. Here are some tips from the Los Angeles Police Department:

- *Keep your laptop with you when traveling.* Never leave it behind, even if you're just standing up to throw something away. Thieves are everywhere and often look just as businesslike and trustworthy as you do.
- *Don't fall for the "distraction technique."* Thieves often work in teams and develop elaborate schemes to catch a business traveler off guard. One common trick is for a thief to use your name and pretend to be an old acquaintance. As you search your memory trying to remember where you met him, a buddy will be making off with your laptop. Beware of anyone or anything that calls your attention away from your computer or luggage.
- *Label your laptop.* Stuff the carrier full of business cards. That way, if your laptop is merely lost, whoever recovers it can find you easily. And if it's stolen, the cops might catch the thief before he or she has discovered and disposed of the business cards.

STRATEGIC ALLIANCES

Don't Go It Alone

If we build it, will they come? That's the question many small-business owners ask themselves before they build a Web site. Sure, some companies find that luring Web travelers is easy, but what if your product isn't sexy?

Elliott Rabin, president of Ridout Plastics, a $7-million company in San Diego, decided to build a Web site. But he wanted to make sure it would generate traffic—and business—for his company, which manufactures plastic components.

Rabin decided to **partner with an existing Web site**, one that had an established traffic pattern. He chose the *San Diego Source*, an electronic financial newspaper. In addition to a $2,500 startup fee, Rabin pays the *Source* $500 a month to maintain his site on its server. The newspaper's site is a hot spot for locals. Many of them find themselves going straight to Ridout's site through the *Source*'s technology page, which Ridout sponsors.

Rabin spends about five hours a week searching the Web for other sites to link with. When he finds a good fit, he e-mails the organization, asks its people to check out Ridout's site, and suggests they consider linking to it. So far he's established links with several universities, including Cornell, which has a plastics department.

Rabin has made Ridout's site *the* plastics site. "We're known as the plastics experts," he says. Sales have increased 50% for one product line alone.

STRATEGIC ALLIANCES

Tap into the New Networking

If you faithfully follow every management trend, chances are you're now striving to be "agile" and "networked," reexamining your relationships with customers, suppliers, peers, and competitors, and devising ways to cooperate and compete to bring about mutual benefits. That's **the 1990s way to think about strategic alliances**. Check out these books for help:

Jessica Lipnack and Jeffrey Stamps have written a trilogy of books about networking. Networking is defined as the process of crossing traditional boundaries to share information and link people and organizations. The best of their three books, *The TeamNet Factor: Bringing the Power of Boundary Crossing into the Heart of Your Business* (Wiley & Sons, 800-225-5945, $29.95), offers many examples specific to small companies.

Cooperate to Compete: Building Agile Business Relationships, by Kenneth Preiss, Steven L. Goldman, and Roger N. Nagel (Van Nostrand Reinhold, 212-254-3232, $24.95), is rich with examples of large and small companies that put the concept of "agility" to use. It's a good read, by credible authors who have spent time in the trenches. All three authors are associated with the Agility Forum, in Bethlehem, Pa., a nonprofit consulting, research, education, and training organization.

Getting Partnering Right: How Market Leaders Are Creating Long-Term Competitive Advantage, by Neil Rackham, Lawrence Friedman, and Richard Ruff (McGraw-Hill, 800-262-7429, $22.95), includes a particularly informative chapter on partnering with other suppliers. The book's suggested guidelines for suppliers—including tips on defining real market value around customer needs—will also help partners stay on the right side of antitrust laws.

• OPERATIONS •

STRATEGIC ALLIANCES

Share Data with Your Suppliers

Some growing companies are extending their data sharing with their large customers. And, rather than having to establish systems that are compatible with their partners' systems, they access the systems directly and use them as if they were themselves employees of the partners. That often requires less technical know-how than does standard electronic data interchange.

G&F Industries, a $22-million maker of plastic molding in Sturbridge, Mass., got the idea for **deep data sharing** from its largest customer. Some time ago, Bose Corp., in Framingham, Mass., asked G&F to put one of its employees on site at a Bose plant full-time. Bose wanted a representative from G&F to be part of a just-in-time manufacturing-resource-planning tactic.

At the time, the notion of an on-site supplier was considered radical, but until electronic communication made face-to-face communication unnecessary, G&F planner John Argitis spent part of almost every day at Bose. He has access to all information available to any Bose buyers, and he has the authority to place purchase orders with G&F on Bose's behalf.

Argitis regularly logs onto Bose's internal network and calls up the plant's materials, requirements, and planning report. The report predicts Bose's inventory needs for the next six weeks. G&F uses the reports to plan its production and to ship products that go straight into Bose's production process, eliminating Bose's need to stock G&F products.

Move In with Your Customers

If your technology isn't getting you quite close enough to your customers, consider sharing physical space with them. According to Ian Morrison, author of *The Second Curve: Managing the Velocity of Change* (Ballantine Books, 800-733-3000, $25) and former president of the Institute for the Future, in Menlo Park, Calif., many companies are **moving in with customers or sharing technology infrastructure with them**.

Many companies in the temporary-help business, for example, are "co-locating" with large clients—literally setting up on site with their partners. Interim Services, based in Fort Lauderdale, Fla., places both professional and clerical temporary workers. It has been opening branch offices on the premises of its large customers. As a result, Interim is discovering all sorts of ways to add value to its services. Because it has an on-site facility at 3Com, a maker of networking solutions, for instance, Interim is now responsible for recruiting, screening, interviewing, and hiring for such areas as shipping and receiving, mail services, and reception.

Other examples of Trojan-horse relationships are travel agencies that develop a presence within large corporate customers, McDonald's restaurants that locate inside hospitals, and Wal-Mart stores that embed their electronic data interchange systems within their vendors' systems.

STRATEGIC ALLIANCES

Clients, Partners, and Everyone Else: Let's Talk

The Web sites that really succeed are those that are different from others in their industry. For example, the corporate-facility construction business doesn't seem rich with online business opportunities, but DPR Construction, in Redwood City, Calif., found otherwise. The $600,000 **contractor created a section on its Web site and called it the Project Collaboration Center**. Behind a whimsical "Admit One" icon resides a password-protected area where the company handles most aspects of online project management for selected clients.

The idea is to give a project's participants—customers, architects, engineers, subcontractors, and even licensing agencies—online capability to share drawings, estimates, schedules, and other work flow-related materials.

And, whenever an unanticipated problem arises at a construction site ("What's this steel beam doing in the air vent?"), workers simply post a digital photo. It doesn't take long for the correct party to answer such questions.

VI

"Each time you
toss out a
'singing' greeting card,
you are disposing
of more computing power
than existed
in the entire world
before 1950."

PAUL SAFFO
futurist, as quoted in *I.D.* magazine

HARDWARE

Together in Technology

Technology purchases can price a small start-up right out of business, but without those high-tech tools, most start-ups can't expect to compete in the global market. Is there an affordable solution? Consider banding with other start-ups and small businesses to **share expensive technology hardware**. You might also share the cost of hiring someone to manage the systems.

Sharing, it turns out, is one of the secrets behind Idealab, a start-up factory in Pasadena, Calif. Founder Bill Gross gathers Internet start-ups and plants them in his facility, where he runs office operations. Idealab companies share state-of-the-art equipment and services, including servers, high-speed digital phone lines, health plans, legal and accounting services, travel planning, and focus-group facilities. "I'm trying to factor out all the common business problems," Gross says. "But I leave in the individual companies just those things that are uniquely related to their businesses."

HARDWARE

Employees Pitch In on PC Purchases

If you're bootstrapping a start-up, you might turn to your employees for financial assistance. **Ask your employees to purchase their own computers**. Then, they can lease them to the company. The company expends less capital, and employees get to buy computers on the cheap.

When PTI Environmental Services, in Bellevue, Wash., was starting out, it needed 15 office computers, but CEO Marc Lorenzen could barely afford half that many. He told his employees that if at least eight of them would buy their own desktop machines, the company would lease them for a year at 50% of the purchase price, which it would pay out monthly. There was only one caveat: He told them not to pick anything *too* fancy.

The tactic worked and helped ease PTI over its initial capital hump. By its second year, cash flow was healthy enough for PTI to buy its own computers. Nevertheless, a few employees refused to switch, volunteering to stay with their own computers at no further charge to the company. Of course, because the employees had been stockholders since early in the company's history, the wisdom of pitching in was clear to everyone.

HARDWARE

Need a No-Nonsense Purchasing Guide?

Once you're familiar with computers and computer-speak, put that knowledge to work. *How to Computerize Your Small Business* (John Wiley & Sons, 800-225-5945, $17.95) is as straightforward as its name implies. Authors Lori Xiradis-Aberle and Craig L. Aberle offer **sensible advice on such basics as how to select software and hardware**, where to buy the goods, how to pay for them, and even how to organize files on your hard disk once the system is set up.

- *How do you figure out what your system should do?*
  Analyze your business and create a wish list based on your conclusions, making sure to include those tasks most likely to benefit from a high-tech makeover. If you have one employee, and it takes five minutes to write one payroll check, it does not make sense, the authors say, to automate that function.
- *Once you've selected your system, how do you budget for it?*
  Factor into your calculations nine categories of expense, including the cost of training, data entry, and insurance.
- *How do you evaluate a service contract?*
  Consider cost, whether the machine is repaired on- or off-site, and whether the vendor can supply a loaner.

How to Computerize Your Small Business does focus on retail operations, so if you're interested in point-of-sale software, bar-code readers, and inventory-management tools, then you're certainly covered. And there's a bonus: The book contains seven sample configurations of hardware, software, and peripherals that are designed for various businesses.

HARDWARE

Learn the Jargon

The first thing you'll need to become computer-literate is a grounding in technical vocabulary—not just what the words mean, but how they relate to your business. For that, try Peter G. W. Keen's *Every Manager's Guide to Information Technology: A Glossary of Key Terms and Concepts for Today's Business Leader* (Harvard Business School Press, 800-262-7429, $18.95). Keen maintains that businesspeople should be as familiar with the essentials of information technology as they are with the principles of, say, accounting or marketing. Toward that end, his book focuses on **200 technology terms and concepts**.

Everyone knows that the computer industry is continually spinning out new products and processes, and more often than not, they are labeled in cryptic three- and four-letter shorthand. To help make sense of what the technology jargon represents, Keen proffers three questions businesspeople should ask: What is its significance for our architecture? What business opportunity does it represent? Does it require me to rethink any aspect of my business plans?

Keen aims to provide readers with tools for evaluating the next big thing, the thing after that, and the thing after that. That's more important than being able to use "TCP/IP" in polite conversation. But don't worry, Keen's book shows how to do that, too.

HARDWARE

...and a PC in Every Home

Employees who use computers off the job are more likely to use their computers efficiently at work. At least that was the thinking at Multiplex, in Ballwin, Mo. To encourage computer literacy, the beverage-dispenser manufacturer **offered low-cost, interest-free financing for employees' home computers**.

The deal cost Multiplex little. When the $28-million company bought equipment for its offices, it added employee orders, passing along the bulk discount to employees. They paid 10% down and covered the balance with monthly installments deducted directly from their paychecks. The only cost to Multiplex was the finance charge on the money it fronted employees for the machines.

It was worth it, says chairman J. W. Kisling. Of the company's 65 office employees, more than half made purchases. Kisling says that he's convinced the program not only improved technical skills inside the company but also boosted employee morale.

HARDWARE

If You Earn It, It's Yours

Tired of buying hardware gadgets for employees? Here's one way that almost **makes the new toys pay for themselves**.

Pat Kelly, CEO of PSS/World Medical, in Jacksonville, Fla., wanted to encourage his medical-instrument salespeople. To get them to sell a load of examination tables within 60 days, he gave each of them the sales incentive up front. He bought every rep a cellular phone, and he told all of them that they could keep their phones only if, as a team, they sold enough. As he was announcing the sales challenge, technicians were installing the phones. "Once they got used to the phones, I knew they wouldn't want to turn them back in," says Kelly.

Each of the 12 sales offices had to sell one $4,000 table for each salesperson in the office. If an office fell one table short, its salespeople would have to chip in and buy one cellular phone themselves. Enthusiasm won out, though, and the incentive did the trick: The sales force not only reached its goal of 70 tables, but surpassed it by selling 105.

HARDWARE

And the Winner Is...

When everyone at your company wants a new computer, how do you **allocate your limited technology budget**? Seko Worldwide, in Elk Grove Village, Ill., has a novel way of deciding who gets new computers. The company runs an essay contest. Seko, a freight-forwarding company, asked 80 of its independent sales reps to compete for 20 fully loaded laptop computers. The contest worked because the topic of the essay was emphatically apropos: Why should you get a new laptop, and how will it benefit you and Seko?

"We wanted to make sure that we were giving the computers to the right people," explains Cathy Moran, Seko's director of sales training and support. "We got a lot of very creative responses." One rep gave the laptop a name and told a story from its perspective. Another even recorded a song, to the tune of the *Beverly Hillbillies* theme song: "This is a story about a man named Bob whose computer was so slow he could hardly do his job..." Top 40 material it wasn't, but it earned Bob a new laptop.

HARDWARE

How to Configure Your Network Server

Everyone knows the easy answer to the question of speed: You can never have too much. But **how much speed, memory, and hard-disk space do you really need**? All three contribute to how smoothly and efficiently your network runs. "Considering anything slower than a Pentium Pro processor is a waste of time," says David Thor, a consultant with Sherwood Research, in Wellesley, Mass. The money saved on a slower processor is negligible compared to the frustration you'll feel if your network isn't fast enough.

A small-business server should have a minimum of 32 megabytes of random access memory (RAM). But you should consider that number as no more than a starting point. Be sure to review a list of your business software needs before deciding how much memory you need. At the high end, "compute-intensive" companies may need 100 megabytes of RAM or more, while "communications-heavy" companies typically need external e-mail, Internet access, groupware, and about 50 megabytes of RAM.

If your company's needs don't match either of those descriptions, consider starting with 32 megabytes of RAM and buying more later. As for estimating how much disk space you'll need, you should go for nothing smaller than a 3-gigabyte or 4-gigabyte hard drive. Some servers aimed at the small-business market come with 6-gigabyte hard drives. And remember, you can always add more hard-disk space.

HARDWARE

Will Peer-to-Peer Serve You?

As soon as you decide to set up a computer network, you'll have to determine if you want a peer-to-peer or client/server network. In a peer-to-peer network, computers are linked together with no central repository for applications; computers communicate with one another to share files.

In a client/server network, servers function as the "nerve center," where such shared applications as databases and e-mail programs usually reside. Clients, or personal computers, "talk" to the server when they need to use the applications. Keeping those programs on a server helps free up the clients' memory and disk space.

For companies with limited computing needs, peer-to-peer networks are a realistic option. But many experts agree that once a company's network has to support 10 or more users, the **cost-effectiveness of peer-to-peer networks starts to decline**. At that point, access to applications can become frustratingly slow. But, experts point out that a peer-to-peer setup may be inadequate in companies where only four or five users must share a very large database.

HARDWARE

Before You Serve Up a Server

As major computer manufacturers aggressively target the small-business server market, you have many server options, whether you're running a computer network or hosting a Web site. According to projections by Sherwood Research, a technology research firm in Wellesley, Mass., small-business-server shipments are doubling annually. While the big vendors rush to fill the growing niche, small companies will benefit from the falling prices and heightened capabilities of today's servers.

But even though servers are getting cheaper and less complicated, they're still not exactly plug-and-play systems. Figuring out how much memory you need and what kind of technical support you should expect can be downright confusing.

Before you purchase your next server, it's a good idea to **forget about the hardware until you've considered the applications**. "The important thing is to figure out what the heck you want to do with the machine and what kind of software you'll need," explains Steven Lee of Random Access Data Systems, a computer consulting firm in Needham, Mass. Sit down with the computer users in your company and discuss their software and communications needs. The list you assemble will help you decide how powerful your server needs to be.

HARDWARE

Buyer, Be Educated

With the vast and confusing array of technology choices, it's easy to see how some companies whiz through several incarnations of phone systems or software before settling down with the right stuff. And even the right choice may be the wrong choice the following year. *Business Consumer Guide* can help you **sort through the bewildering options**, guiding you to make well-founded purchasing decisions.

The publication, based in Watertown, Mass., accepts no advertising. In its issues, you can find thorough reviews of printers, CD-ROM drives, and the like as well as details on the availability of such technical appointments as modem jacks and fax machines in hotels. One analysis of overnight-mail delivery services included a review of the tracking software many such companies now offer.

Because the *Guide* assumes that its readers have no prior technical knowledge, each issue includes a customizable decision-tree chart that can help small companies take a systematic approach to their technology acquisitions. A year's subscription of 12 monthly issues costs $129, and it is available by calling 800-938-0088.

HARDWARE

Get the Most from Firewall Power

Imagine that your house is surrounded by a huge metal fortress. It has one door, at which a gatekeeper stands and intercepts visitors. He asks them where they're from and then decides whether to let them in based on your criteria. Furthermore, the gatekeeper permits people to leave your house only if they are headed to preapproved destinations. A firewall performs comparable tasks to guard your computer network.

Firewalls are programs that act as gatekeepers between your computer and the world. Different types offer different levels of security. Bluestone Consulting, in Mount Laurel, N.J., relies on a firewall not only for security and protection, but also for the detailed reports it generates. Steve Haas, manager of information systems at the software company, checks hourly reports that track such "suspicious" activity as repeated attempts to log in with bad passwords. He reads daily reports of all traffic that flows over the $16-million company's modem line.

"It's worth it to buy a separate machine just for the firewall software," says Haas. Allocating processing power from another machine on the network could slow things down because everything that the company sends or receives through a modem goes through the firewall first.

Remember that firewalls vary in ease of implementation and maintenance, with the most secure requiring the most work. Prices range from $3,000 to $100,000, so it's important to do a little homework. You don't want to buy a firewall that does more than your company needs.

HARDWARE

Track Hardware Depreciation

Most electronic equipment—e.g., computers, fax machines, telephones—has a five-year life on tax depreciation schedules. After that, it's considered worthless. Although it's a lot harder to keep tabs on a phone than a truck, **keeping track of your small equipment can pay off at tax time**.

Accountants advise tracking individual pieces of equipment from the day they're purchased. Why? Suppose a company buys 10 PCs at different times over the years and that three of them are tossed for upgrades. If the company hasn't tracked the individual machines, it will defer realizing the full value of the depreciation on the three that were tossed and will have to wait until the entire class is depreciated.

A tracking system also pays off should a machine disappear. A company that takes inventory regularly will realize when a laptop is missing, and it will be positioned to deduct its undepreciated value. A simple way to do it is to put bar codes on all your hardware and scan the numbers into a computer spreadsheet program.

HARDWARE

Telecom Shopping Tips for Soloists

Many business founders are shocked when they suddenly find themselves in charge of MIS, having to decide which computers, office machines, and telephones they should buy. The world of telecommunications can prove a minefield for the uninitiated, and learning a bit about it can make the difference between finding good, inexpensive service and being burdened with overpriced, useless features.

Fortunately, the small-office/home-office market is hot. And people are writing how-to books geared toward the soloist. June Langhoff's *Telecom Made Easy* (Aegis Publishing Group, 401-849-4200, $19.95), for example, deserves praise for its **thorough coverage of telephones for home and small businesses**. Now in its third edition, it includes chapters on modems, cellular phones, voice mail, phone systems, and phone bills. It defines terminology and offers checklists aimed at helping users decide which products and services are right for them. Especially helpful for do-it-yourselfers, the book provides tips on troubleshooting and repairs. Just leafing through it will get you started, preparing you for smarter dealings with salespeople or the telephone company.

• PURCHASING •

HARDWARE

Let a LAN Leverage Productivity

United Science Industries, in Woodlawn, Ill., was growing so fast that it garnered a slot on the 1995 *Inc.* 500, but CEO Jay Koch was dismayed that labor overhead was eating up company profits. The general contractor had no computer network, so employees were duplicating one another's work. For example, the accounting department had to gather and enter data from field managers' handwritten sheets to generate invoices, a process that took at least two months.

Koch knew he needed networking power. A consultant recommended a local area network that cost about $76,000. Wiring 15 networked computers cost $8,000, and Novell NetWare for 50 users and the server totaled $11,400. Koch decided to dump his 10 Macintosh computers for PCs. Replacement computers and software cost $47,500, and he paid another $9,000 for in-house training.

"With the network, people spend **less time hunting for information and more time working** on tasks crucial to operations," Koch says. Less paperwork is floating around the office, since all workers can gain access to the same information from their computers.

Accounting has access to field managers' files, so invoicing takes only 10 to 12 days. Koch estimates that he's accelerated accounts receivable by 5 to 10 days and lowered labor costs by $20,000 a month.

HARDWARE

Are Your Employees Wasting Time?

Looking for an excuse to dump the company fax machine in favor of fax/modem software? Saving employee time may be the answer.

At Roddy Temporary Services, a $2.3-million temp agency in Ann Arbor, Mich., employees fax information among three offices and to about 20 customers. On any given day, says CEO Philip Roddy, each of his nine employees spends a total of **one hour at the fax machine**. And Roddy's situation is far from unique.

In fact, most employees spend 30 minutes a day at the fax machine, according to a survey of 200 small companies by Impulse Research, in Culver City, Calif. Each person spends about 24 minutes a day at the photocopier, and printing takes up 21 minutes. The companies surveyed averaged nine employees each, of which four visited printers, faxes, and copiers about nine times daily—an average of 75 minutes total. That's a staff total of 25 person-hours a week, or the equivalent of a part-time employee.

• PURCHASING •

HARDWARE

Finding Cheap Voice Mail

A new voice-mail system can cost you thousands of dollars. After all, you need to install new telephones, wiring, and software. Ray Smith, president of WinterBrook Beverage Group, in Bellevue, Wash., found a way around those expenses.

The $36-million company was subletting office space from another company that suddenly decided to move. Rather than uprooting its telephones, the **departing company offered to sell the phones and its attendant voice-mail system** to WinterBrook. Smith says his company paid about $1,000 for a very serviceable system. The automated-attendant feature offered incoming callers four options—press 0 for operator, 2 for company address and fax number, 3 for employee directory, and 1 for all other options.

It's true, of course, that most used systems won't have all the up-to-the-minute features and functionality. Smith acknowledged that his lacked speed and had minimal message-storing space for the 20-employee company. Still, he appreciates that not only did the system come with a bargain-basement price tag, it also reduced the company's personnel costs.

HARDWARE

How to Choose the Best LAN

For Erica and Brian Swerdlow, selecting a new network operating system for EBS Public Relations, in Northbrook, Ill., was a trial. "It was making us crazy," Erica recalls.

When they founded their company in 1993, the Swerdlows set up a local area network (LAN) so all three employees could share data. The company was too small to need a client/server network, so the Swerdlows linked the PCs into a peer-to-peer network.

A few years later, with eight employees and sales of $500,000, the Swerdlows decided it was time to move the company from its basement start-up office and install a more sophisticated network operating system.

EBS turned to a systems-integration company, and it recommended Novell NetWare. Because the upgrade would be costly, the Swerdlows sought a second opinion. That consultant gave an endorsement no less firm—Microsoft Windows NT Server. Thus began an ordeal that lasted several months. The Swerdlows read computer magazines, and they talked with reporters, computer industry clients, and clients of clients.

Their conclusion: While **there is no single "right" answer to selecting technology**, everyone wants to convince you that his or hers is exactly the right solution. "You've got to gather all the information," says Brian. "Then form your own opinion." The Swerdlows, whose company now employs 16, chose the Novell system.

HARDWARE

Give Old Computers New Life

Have you finally upgraded your computer system? **What can you do with the old hardware?** Your old computers are likely so outdated that nobody will want to buy them. But they might still have a little life left. Here are some options:

- *Upgrade an old computer by replacing its CPU chip.* Many hardware manufacturers offer kits that come with replacement chips you can install yourself. For less than $100 you can upgrade a 386 to a 486, and you might eke out one more year of service. That machine can handle single-application software tasks. By not networking the computer, you can extend its capabilities even further.
- *Turn the computer into a dedicated online machine.* Adding a modem can transform an old dinosaur into a host for a customer bulletin-board service or an online workstation for employees. Randy Fields, CEO of Park City Group, in Park City, Utah, has one dedicated computer with Internet access for his 130 employees. Because the machine stands alone, hackers can't reach the company's network.
- *Donate it to charity or a nearby school.* This way you get a limited tax deduction.
- *Sell it to a salvage company* that reclaims such useful parts as the gold plating and recycles the rest.

HARDWARE

Optimizing Your Electronic Organizer

The right personal information manager (PIM) is an invaluable tool. The computerized datebooks, to-do lists, telephone directories, and memo records can be great for tracking one's time. Or they can be a huge time waster. Because PIMs vary so much in look, feel, and operating philosophy, it's important to **audition any PIM you might buy**. With a candidate PIM in hand, consider the following checklist:

- Decide whether you want a program that defines how information will be organized or one that lets you set up the structure you want. If you plan to use the PIM for long-range planning and strategic overviews, look for programs with as little built-in structure as possible.
- Choose a PIM that requires you to enter, say, name-and-address information only once. It should automatically link addresses and phone numbers to appointments in the schedule book and to your e-mail.
- Test the PIM's data-recovery system, and lean toward models that provide ways to extract data from corrupted files. The more you enjoy working with a program, the sooner it will become the repository of your most important business information. And if you travel with it, you don't want to be stuck on the road with a data failure.
- Check how hard or easy it is to move data from online sources into the PIM. Clumsy cut-and-paste protocols or limitations on file length turn your PIM into a big disappointment.

SOFTWARE

Are You Ready for the Turn of the Century?

You've heard about the Year 2000 scare—software that uses two-digit date entry fields (dd/mm/yy) might struggle with "00" and stymie your system. No matter what your software vendor says about your system's adequacy to make the transition, it's still smart to **test everything that's critical for Year 2000 compliance**. Try putting your software applications through their paces using a variety of date combinations falling before and after January 1, 2000.

For example, an invoice application might use three dates: date sent, date due, and date paid. First, set all three dates in 1999. Then use 1999 as the sent date and due date, but make the paid date 2000. You'll have to do a separate test to be sure that the software can handle February 29, 2000, because—as if this all weren't entertaining enough—2000 is also a leap year.

Peter de Jager, a Year 2000 consultant based in Brampton, Ont., recommends having professionals do the testing "because they have the goal of making the system fail." It is also imperative to back up everything or, ideally, to run the tests on a stand-alone machine. "Sometimes when things go wrong, they go wrong in spectacular ways," de Jager warns.

SOFTWARE

Your Word Processor Can Do Double Duty

Don't bother spending money on an expensive customer-tracking database—you may already have the software you need right on your computer. Most high-end **word processing programs also offer stripped-down database capabilities**.

At Finagle-A-Bagel, in Boston, founder Larry Smith used to save each week's 15 or so customer suggestions and complaints on Post-It Notes. But the $12-million bagel business was expanding rapidly, adding several new stores. Smith decided that it was time for a database. His information-systems manager, Cosmo Nardella, turned to Microsoft Word for help.

Nardella built a new database in one hour. Every time a customer calls, his or her name, phone number, and comments go into the database. If a customer calls again, a customer-service rep can retrieve the data and talk to the customer, conversing as if every detail of their last conversation were still fresh in mind.

Managers also sort through the database to identify problems and fix them. Smith retrains or disciplines store employees whose names repeatedly turn up in complaints, and he gives careful consideration to comments about the quality of his bagels.

SOFTWARE

Software As You Like It

When you decide to add or upgrade a customer-service database, don't forget that you can have it configured any way you want. Harvey Katz built the reputation of his $2.5-million independent hardware store on personalized customer service.

When his store, Harvey's Hardware, in Needham, Mass., finally computerized in 1994, Katz found that he and his well-trained staff **didn't need three-quarters of the built-in features** of the system, which was specially designed for hardware shops. Why have a computer tell you when to reorder duct tape when any of the 22 people who work in the store can tell you when you're down to your last few rolls?

When he relied on paper systems, Katz had always made a habit of filing his customer information by phone number rather than by name. Because he has a poor memory for names, he had the software vendor change the system when the company recently upgraded the database, adapting it to a phone-number-based filing method. That method is now a standard feature on the latest release of the software.

SOFTWARE

Plan Before Anyone Programs

What are the keys to successful software development and office automation? It's crucial to find the right person to oversee the job and for you to give the project the level of attention you'd give to any other core operation. Consultants emphasize the importance of defining the problem you want solved and planning every detail of the solution.

"There's a rule that professionals go by," says independent consultant Gordon MacDonald, who operates out of Shawnee, Kans. "For every hour you spend on planning—meaning not touching a computer—you save five hours in implementation. **Before anybody gets to write software, create mock-ups of all the screens** the program is going to produce, with the menus and options. Develop a flow chart explaining, in plain English, how the program is going to proceed, so that nontechnical people can see how it will react to specific situations.

"And there should be a clear statement about what the program will and won't do," adds MacDonald. "The kiss of death is when a programmer spends an hour with an end user and then sits down to write."

SOFTWARE

Keep Custom Software on Track

What are the most **commonly asked questions about getting custom software**? Custom-software developers Bill Zimmerman and Dan Johnson of Pro Systems, in Gurnee, Ill., say people are most concerned about costs and changes:

How should people prepare for meetings with the developer? We want to know about the business and the goals of the project. If there are people who understand the operations around the software better than the CEO does, have them at the meeting.

What are the standard pricing and payment terms? The best pricing has aspects of both fixed-cost contracts and time-and-materials contracts. Using fixed cost for the main project gives the developer an incentive to understand exactly what you want. The time-and-materials approach works well for additional work before and after. Some time-and-materials projects have penalties for missed deadlines. With fixed-price agreements, we bill 25% down, 50% in progress, and 25% upon completion.

How should one make changes along the way? The earlier the better. You need to recognize that a change may cost extra if the contract stipulated that things were going to be done a different way.

What's the best way to contain costs when a project starts to go over budget? By that time, it's too late. Before work starts, tell the consultant, "I want to monitor costs. What intermediate points are there so I can tell if we're on schedule?" If there's a cost overrun, it's because somebody didn't realize the complexity or scope of what was actually needed. If it's the consultant's fault, you may want a third opinion on what's being done.

SOFTWARE

Buy It or Build It?

Tim Litle's business processed mail-order companies' credit-card orders. Each day, Litle & Co., in Salem, N.H., faxed its customers information on the previous day's credit-card transactions. Business really took off, and the company got so busy that Litle had two people who spent all their time manually feeding a fax machine. It was clear to him that it was time to automate.

Litle arranged for a custom solution, which took three person-months of in-house programming and $30,000 in new equipment. But, because the custom-made system was unacceptably slow, Litle soon began looking for its replacement. A project team, under the direction of operations vice-president George White, was able to swap the proprietary slowpoke for a $20,000 turnkey fax server, which has delivered great savings.

The company has since been acquired and is now called Paymentech, but the moral of the story, according to White, still holds true: **"If you can buy it, don't build it."**

SOFTWARE

Custom Software Can Kill You

"For years, I wanted everything in my company done perfectly," says Jeffrey Mount, president of Wright's Gourmet House, in Tampa. But Mount got a little carried away when he started searching for a point-of-sale (POS) software program to automate order processing at the $2.4-million restaurant.

"Knowing that I couldn't buy the perfect system, I decided to **hire a programmer to write the perfect program. What a mistake!** For two years we muddled through development hell. The software company had no POS experience and no food-service experience. Its lead programmer didn't even visit our facility until six months into the coding. By then, the (wrong) foundation had been laid. We spent the next couple of years—and nearly $70,000—trying to clean up the mess. One day the software-company president dropped by and said that if he knew at the start what he knew now, the firm could have done a much better job. Of course, if I would be willing to pay to start over, the company would give it another shot."

Instead, Mount reassessed his options. Armed with a list of what he wanted the software to do, he attended an industry trade show and checked out off-the-shelf software. "The company promised to modify the system to fit our needs," Mount says. "Is it perfect? Of course not. Have we had problems? You bet. But for the most part, the system gets the job done beautifully. My life is much saner, and my customers and staff are much happier."

SOFTWARE

Run Ahead to Stay Ahead

Technology moves so quickly that a competitive edge can turn dull overnight. That's what Gary Haselton learned when he was running Haselton Construction. Haselton customized an off-the-shelf spreadsheet program to perform project estimates. His goal was to get more accurate estimates and turn them around faster. But soon after he finished, technological advances erased his competitive advantage.

"I spent 700 hours writing an estimating program, at a cost of $10,500 at my $15-an-hour contracting rate," he says. "But within six months, there was a commercial program on the market for $50 that was 10 times more powerful. That was a lesson to me in how fast technology moves and **how quickly innovations become obsolete**."

Haselton, who now runs EPIC Multimedia, in South Burlington, Vt., did, however, manage to make his software pay off. The program he wrote worked better for him than the commercial software, and he was soon able to give clients estimates that were accurate within 2%. A typical builder's accuracy is about 10%.

"We computerized on a 'time shoestring,' buying hardware and software only when not having it would mean losing orders or something equally disastrous. A crucial part of the strategy was to select products as much for the way they fit into our schedule—time needed to learn, for example—as for their features and price."

ALLISON ROSSON
owner-president of Mike & Ally,
New York, N.Y.

SOFTWARE

Overruns: The Only Guarantee

Managers at small companies agree on one aspect of **custom programming: It will take more time and money than you expect**. When Skyline Displays, in Burnsville, Minn., set out to automate the designing and order processing of trade-show booths, it took an entire year longer than management had anticipated just to get the software running properly. And please, company insiders say, don't even ask about the cost overruns.

Dale Uhl of Wastren, in Idaho Falls, Idaho, discovered that he and the staff of his $9.6-million waste-management business were giving the custom programmers mixed signals about what they wanted the software to do. No wonder, he says, the project took much longer than it should have. In fact, the entire project was an unmanaged waste of time. Uhl says that even though he persevered, the software never worked properly, and he had no choice but to abandon it.

SOFTWARE

Systems vs. Problem Solving

Are you thinking of commissioning custom software? John Thackara of Netherlands Design Institute, in Amsterdam, has some advice for you:

"When we started up in 1993, we were determined to keep our operation small. So we turned to technology—internal networks, a shared database, teleconferencing, e-mail, online discussion software, and the World Wide Web. It was in the context of a business that did not yet exist that we set out to build the systems.

"That was blunder number one. We ended up **separating the systems from the questions** that should have defined them, namely: What is our business? How can we do it better using technology? What software and equipment does a given project require?

"I asked an outside expert to set it all up. That was blunder number two. The result: Nobody in-house really knew how the systems worked.

"We had to rush at the beginning because the builders were on-site. So we ordered cabling, computers, a $50,000 telephone system, ISDN connectivity, and boxes and boxes of software. The consultant had to guess, from hurried conversations with me, what the system was supposed to do. We ended up with a fine-looking front end to a system whose inner architecture none of us understood.

"I suspect we lost several person-years of productive time. We should have asked the right questions at the beginning. Today, staff members 'own' bits of the information environment. We don't spend a penny on new software or equipment unless a project demands it. Are the lessons we learned worth the cost of the blunders? You bet they are."

UPGRADES

Why Everyone Needs to Agree

When Robert Jacobson was president of a virtual-reality design studio in Seattle, he involved the entire company in technology purchasing and upgrade decisions. He swears by this purchase-by-committee process. Although it was occasionally time-consuming, everyone understood and supported the new acquisitions.

"We had only five employees, so **we passed the responsibility for technology decisions around**, depending on the particular need. For example, when we networked our computers, at least three of us had opinions on the subject, so we met and discussed them. First we compared the technical capabilities and prices of the different systems. Our vice-president for applications and development listened to all the opinions. Once we had general agreement on what type of system we needed, he took all the information and made the purchase, because he has a background in networks.

"We never had one person acting as 'technology czar.' If you don't make decisions by consensus, people become unhappy. If they went out and bought their own equipment, that would lead to chaos. For instance, when we bought word-processing software, we all agreed that it had to be a package with which we were familiar and that was readily available and affordable. Then someone who had time bought the package."

UPGRADES

Technology Tester Saves Time

Like many other techies, Steven Ettridge, CEO of Temps & Co., based in Washington, D.C., is forever replacing his tools with newer, better models. For example, he recently switched personal digital assistants, jettisoning his Sony Magic Link in favor of the PalmPilot. On the road, Ettridge uses the PalmPilot's e-mail feature to read messages in the car (or occasionally during a plodding meeting). Ettridge also relies on a mobile telephone and pager.

How does he **keep up with all the new models and features**? He doesn't. He has delegated that job to his systems manager, who serves as a technology coach. "He gets to buy all the productivity tools he wants, and he gives me a *Reader's Digest* report on whether it's worth playing with or not," says Ettridge. The 20% to 30% of the stuff that doesn't pass muster can generally be returned for full refunds within 30 days.

Companies lacking an obvious "toy meister" to serve as technology coach need not despair, says Ettridge. "If you have college kids, one of them would love the job."

UPGRADES

Computerize on a Shoestring

Time is of the essence in Allison Rosson's $1-million design and manufacturing business. Mike & Ally, in New York City, makes decorative purse, bath, and tabletop accessories. Early in the company's history, the founders knew they had to computerize, but didn't have the time to research the equipment. "There was no question that the big bang approach—closing up shop for a technology overhaul—was not for us," Rosson says. They decided to go digital at the height of one year's Christmas rush. Taking even one week off to automate would ruin the ship dates for 40 customers—$40,000 worth of orders. Hiring someone was out of the question. After all, the two founders were taking home only $25,000 a year each.

They had no choice but to **computerize on an as-needed basis**: buy hardware and software only when not having it would mean losing orders. A crucial part of the strategy was selecting products as much for the way they fit into the company's schedule as for their features and price. For example, Rosson chose an accounting package based on how easy it would be for her to learn to use it. When a customer demanded electronic data interchange (EDI) capability, Mike & Ally paid a third-party service rather than invest hundreds of hours and staggering sums to bring the process in-house. Rosson dubbed her strategy "computerizing on a time shoestring."

The payoff is evident: The company maintained cash flow while speeding operations enough to expand the business. Since 1995, the company has taken on 900 new customers and added 11 new product lines.

UPGRADES

Ask the People Who Use It

Several years ago, executives at Keystone Helicopter, in West Chester, Pa., had to purchase a software upgrade, and they wanted to do the right thing. Rather than leave the decision up to any one person, a handful of executives at the $30-million aviation service company listened to several sales pitches and chose new companywide software based on what they heard. But the process failed, and the software bombed because it wasn't focused on aviation concerns. After a year and more wasted money than co-owner Peter Wright Jr., will admit to, Wright **turned to employees for help in buying a replacement system**.

Volunteers from each department started by documenting internal processes and asking their peers to suggest improvements. The volunteers met weekly to compare notes. Based on what they found, they interviewed vendors, reviewed software, and recommended a system.

With most of the employees pitching in, the company ended up with effective software—and some improved internal processes.

UPGRADES

To Upgrade or Not to Upgrade?

When to upgrade and when to wait? That question haunts many CEOs, especially when a major new release like Windows 95 hits the market. That release, like many new software releases, was plagued by missed deadlines and buggy early versions. No wonder many CEOs are wary about upgrading as soon as a new version appears on the scene.

"If you catch every wave that comes along, you lose your focus on the business. Suddenly, nobody can find files, and you're busy making it all work," says James Bush, chief financial officer of $20-million Priester Supply, a distributor of electrical equipment in Arlington, Tex. "We still have a business to run, whether the latest software comes out or not."

Other small-business owners agree that **waiting for later releases of new software makes sense**. India Hatch, CEO of Taos Valley Resort Association, in Taos, N. Mex., decided to stick with Windows 3.1 during the initial release of Windows 95. For the $1-million reservations bureau, computers are mission-critical. "The computer consulting firm we use recommended that we upgrade right away, only a year after we had bought Windows 3.1," Hatch says. "But for upgrades, I have to look at the cost and what we're going to gain in terms of features and efficiency."

UPGRADES

Of Scoops and Scope

Abbott's Premium Ice Creams, an ice cream distributor in Conway, N.H., needed to upgrade its antiquated information technology. Its stand-alone PC just couldn't support the company's growth and its quickly expanding product line. The computer was so old it had no hard drive, and the company's programs were all on diskettes.

It would have been entirely reasonable for the company simply to buy a Pentium machine with a good accounting package and be done with it. But Abbott's founders, Margaret and Sut Marshall, knew that if they went with a PC server, they would have to upgrade again in only a few years to accommodate the company's growth. So they elected to **move beyond PCs, all the way up to a powerful minicomputer**.

After shopping around, they purchased the IBM AS/400 Advanced Systems Model 200 computer for just under $8,000. The machine required so little maintenance that some functions—such as backing up data each night—run hands-free.

Now, using six terminals scattered throughout the building, all areas of the 25-person company, from accounts receivable to shipping, immediately and automatically share order and other information. "Switching to the AS/400 from our old computer was like going from riding horseback to driving a car," says office manager Nancy Calvert.

UPGRADES

Let Customers Be Your Guide

Tom Smith Industries (TSI), a 17-year-old molder of thermoplastic material in Engelwood, Ohio, made an important upgrading discovery. TSI found that **waiting until customers have upgraded** their computing technology is not only economical, it can be good business. For example, in 1996, when the newer, more capable Windows 95 and Windows NT were being heavily promoted, TSI's PCs were still running Windows 3.1. Windows 3.1 and the software that's written for it still worked with the software of the clients with whom TSI exchanges engineering and administrative files.

"We still used a lot of the older programs that simply wouldn't work with Windows 95," says president Tom Smith Jr. "Our customers used them, and it would have thrown everyone off if we had changed." When the customers finally upgraded in early 1997, TSI followed suit.

UPGRADES

Streamline, Then Automate

Lantech's plant runs on technology that could have been installed 40 years ago. Instead of a shop humming with numerically controlled lathes and automated assembly machines, workers are wielding drill presses and hand tools. In place of computers generating specialized work orders, conspicuous "whiteboards" diagram the assembly process. The managers of Lantech, in Louisville, Ky., use cue cards and strips of tape to track supply levels and production flow.

Founder Pat Lancaster dragged his company from heavy computer use back to the dark ages. Why? He wanted to **step back and get a handle on disorganization**—the computers had actually thrust the company into an era of faster waste. "Previously we were just automating chaos, buying expensive machines to do wasteful things at higher and higher rates of speed.

"We were lying to the computer all the time," Lancaster recalls. "If we had trouble getting deliveries as fast as a customer wanted, we'd tell the computer the order was two weeks older than it actually was. The computer would reschedule that job, but then all the rest of the orders would be held up.

"The numbers the computer was working on were never right. It might say there were 10 items in inventory when we could see there were eight. We'd just enter the right number over the wrong one, which guaranteed that the error would show up again. You could almost never track down where the error occurred."

Stepping back from automation enabled Lancaster and his employees to understand the company processes before moving back to the high-speed world of automation.

UPGRADES

Take Team Route to Tech Decisions

Jos Kleynjans, CEO of Trading and Manufacturing Industries, an $8.5-million strip-door manufacturer in Pittsburgh, explains how he makes purchasing decisions: "If a technology purchase will affect the whole business, then we **make the decision as a group**.

"For example, when we were thinking about automating, we discussed the decision with everyone who would be involved, no matter what position he or she held in the company. Everyone agreed that we were ready. When we invited suppliers in to bid, they made their presentations to the seven department heads, who later informed me of their preferences. Five chose the same supplier.

"I trust my department heads to make technology decisions because most have worked for other companies where they've had to make similar judgments. Often I make a decision in my own mind, but I don't express it until the group reaches a decision. If they don't agree, then I'll challenge them. If they can prove me wrong, I accept their decision."

UPGRADES

Count on Pros Who Know Your Business

Stan Wetherell is the cofounder of the sign franchisor Sign It Quick and co-owner of the $1-million Sign It Quick store in Columbia, S.C. He learned to **take the advice of outside professionals** when making technology purchases or upgrades.

"I make the decisions, but I do make mistakes. I hired a man who had been selling us computers for a long time to put in a computer system for graphics. He's knowledgeable about software, specifically point-of-purchase and accounting packages. Turned out he didn't understand what the requirements were for a graphics computer. I spent $3,800 apiece on two computers he said I should buy, and they weren't as good as the $2,000 computer I got from Best Buy a year later.

"Now when I go outside the company for help in making a technology decision, I talk to several computer experts and make sure I get references from people, preferably those in my industry, who've used the recommended products."

UPGRADES

Relying on Local Resources

Dale Alldredge, president of $8.2-million Technic Tool, a manufacturer of outdoor power equipment in Lewiston, Idaho, relies on **local computer retailers to help guide purchasing decisions**:

"When I decided to put computers on every employee's desk, I called a retailer I'd known socially for 20 years. He sent people over immediately, and they asked us a hundred questions. Then they recommended specific hardware and software and installed it. Even today, we call the store when we have problems, and someone is here within 15 minutes to help us.

"When my accountants recommended an automated accounting system, I called up my retailer. I wanted him to sell it to me so that I could rely on his store for technical support. He researched the software and then ordered some samples so his staff and I could test it. Finally, I bought it from him."

UPGRADES

"Hello, I'm a Human Rep."

Can great technology actually hurt your business? It can if your staff and customers hate it. That's what Harvey Epstein, president of Keena Mfg. Corp., in Newton, Mass., learned about his voice-mail system. A poll of customers and staff showed a nearly unanimous distaste for voice mail, and Epstein realized that he could **upgrade his company's performance by actually downgrading its technology**.

The company's phones are not allowed to ring more than twice, and a live customer-service representative answers every call, quickly dispensing information or taking an order. Epstein has found that the low technology enhances his staff's ability to personalize service. He instructs reps in the art of schmoozing. His reps get to know their customers by asking open-ended questions like, "How's the weather?" or "How was your vacation?" That's something that even the highest-tech voice mail can't do.

UPGRADES

Do You Have Good Reasons to Upgrade?

Job-costing software can simplify all kinds of purchases. Such software is typically used in the construction industry to track costs versus revenues on a per-job basis. But it can have other uses.

Jim Osmundson, owner of Sierra Information Services, a graphic design company in Yacolt, Wash., uses job-costing software to **evaluate whether or not it's time for a software upgrade**. Osmundson uses the tools that come with the QuickBooks Pro accounting package (Intuit, 800-4-INTUIT) to track the time spent and the specific software used on his clients' projects.

At the end of the year, he generates a report showing how many hours he billed for each graphics program used. Before upgrading any software, he checks the report. If he billed fewer than 50 hours during the previous year using a particular package, he doesn't bother upgrading it, saving himself money as well as the time he would have had to spend learning the upgraded software.

UPGRADES

Upgrade on the Cheap

New software comes with its risks. For a small business, specialty software costs can be prohibitively expensive, and if your employees don't easily learn how to use your new software, they may ignore it altogether.

One way to test-drive packages before purchasing and installing them is to **volunteer to beta-test software** that interests you. You'll be able to give your key employees an opportunity to try it before you spring for the big investment. Dan Caulfield, president of Hire Quality, a job-placement firm based in Chicago, took that approach with a tricky database-search software package. Not only did that move prepare his employees for the upgrade, but Caulfield also saved money.

Philip Ellinwood, chief operating officer for Ritchey Design, a $15-million mountain-bike component designer in Redwood City, Calif., struck a similar deal with SBT for its pricey software package called WebTrader. Ritchey Design wanted the software to conduct sales transactions on the company Web site. In exchange for his willingness to test the package and put SBT's logo on his company's Web site, Ellinwood negotiated a software deal he could afford.

Check for Compatibility Before You Upgrade

It can be tough to keep up with software upgrades, especially if your company uses several different programs. New versions appear on the market faster than a small company can afford to buy them. How do you know when it's worth investing in a snazzy new version of an old standard?

Bill Young, of Young & Perry Insurance, in Bridgewater, N.J., waits for the okay from the software vendor that created his primary insurance program. Since he relies on that insurance program, it's essential that **all other software be compatible with the core-business software**. He applied that rule to new versions of Windows. "I've been a pioneer enough times to know that if my primary software vendor isn't comfortable with a new version of some software, then I'd better stay away." Until he gets the word, Young says, he never upgrades.

UPGRADES

Who Says You Need to Upgrade?

When Keena Mfg. Corp., in Newton, Mass., upgraded its hardware, president Harvey Epstein considered taking the traditional route and upgrading the database software, the heart of the company's system, as well. After investigating several database packages with "manuals the size of the Manhattan Yellow Pages," Epstein realized that for all their muscle, the newer packages really could do no more for Keena than the company's trusty PFS: First Choice. That meat-and-potatoes package had been easy for everyone at the company to learn.

Why did Keena decide not to upgrade? The company's largest database was only 11,000 records, and using one of the new programs to manage it would have been like using a bazooka to kill a flea. First Choice allowed the company to do everything it needed, and nobody had to struggle to learn a new system. "Now I'm a hard-core believer that **a blend of old and new technology gives the best return**—in our case, 8% annual growth," Epstein says. "The new should be added only when it's clear that it will bring absolute value to the company."

VENDORS

Don't Pay Until It Works

Buying his first network was a disaster for Joe Alloway, owner and founder of Credit Union Marketing, in Mt. Pleasant, Mich. The idea was to link writers, designers, and typesetters and enhance efficiency. The $4-million business publishes 150 newsletters.

Alloway kicks himself now for his failure to realize that when he bought the local area network (LAN), both he and the vendor, Computerland, were out of their league. At least Alloway had the smarts to **ask for a written performance guarantee** from Computerland. The parties agreed that he would pay one-third of the $27,000 price tag on delivery and the remainder when the network was up and running. He realized just how smart he had been as soon as a slew of problems beset his LAN.

According to Alloway, the network never stayed up for more than a day or two at a time. And whenever it went down, Computerland's solution was to add a new piece of hardware or software—at Alloway's expense.

"It was such a disaster that the business went on hold for a year," says Alloway. In fact, he blames the network problems for an embarrassing printing error (a misplaced decimal point in an annual report) that cost him one of his most important clients. "That was $40,000 a year lost, right there," he says. Thanks to his contract, Alloway was able to withhold payment, finally ending his relationship with Computerland and selecting a new network vendor.

VENDORS

Fie on Foreign Call Costs

International calls can break the bank. But when you're setting up a branch office overseas, evaluating new vendors, or hiring foreign reps, what can you do to ease the pain? One way around prohibitive bills is to **use an overseas callback service**. Because it completely bypasses the local country's phone company, it can save you as much as half the cost of overseas calls.

Karen Hammons, president of Parker Golf International, a wholesaler of customized golf services in Magnolia, Miss., found out about the service from an Internet mailing list she subscribes to. Hammons uses the service for faxes and phone calls, and she pays only 32¢ per minute to call Australia, for example, instead of 62¢ for dialing direct. To use the service, she dials a U.S. number and hangs up. Shortly after, when her phone rings, it's a recording from her overseas service. Hammons dials the overseas number and is connected immediately.

Before Hammons started using the service, Parker Golf's monthly phone bills climbed as high as $800. Now the bill is two-thirds that. And Hammons is also using a third-party service for her broadcast faxes. "I like to use services. That leaves us to focus on our core business and let other people do what they do best. We save money in the long term, considering our time."

Who's Who in High-Tech Savvy?

With countless thousands of people serving the industry, nowhere is the term "consultant" more ambiguous than in the computer field. If you need **outside expertise in office automation**, think in terms of four broad categories:

- *Independent consultants.* In their purest form, these are objective advisers. For the most part, they serve small and midsize companies, and some specialize in particular industries. They don't work cheap. Sometimes they bid on jobs on a flat-fee basis, but most prefer to bill hourly rates. Experienced consultants command from $80 to $200 an hour, depending on their skills and local economic conditions.
- *Value-added resellers (VARs).* These close cousins to independent consultants are authorized reps for software and hardware vendors. They get a piece of the action when they sell products in their channel, so they come with distinct biases. Ethical VARs disclose their relationships, but you may encounter an unscrupulous rep who masquerades as an independent. VARs generate income from two sources: hourly consulting fees and commissions.
- *High-tech temps.* Thousands of people who consider themselves independent consultants work as temps, farmed out by "body shops" for projects or assignments that run for six months to a year or more. High-tech temps provide staffing flexibility and the specific skills needed at the right time.
- *Giant consulting firms.* This is the gold coast of computer consulting, in both cost and prestige. Such names as IBM and Andersen Consulting are among the more familiar. Their services are expensive, with hourly rates running in the $200 to $350 range.

VENDORS

Beware the Guru Effect

Even sophisticated businesspeople with strong negotiating skills can get **burned by expensive computer consultants** who—because their auras demand awe and respect—are never questioned about their systems or implementation.

Not long ago, the partners of a Los Angeles law firm met a consultant who promised to bring them "up to speed" on computers. "He was trying to convince us that all the lawyers needed PCs," recalls a former associate at the firm. "He said we could be the first guys on the block to distribute laptops to our clients, so they could be networked into our system, with passwords to their files. We could be really high-tech and paperless."

The "guru" put a $200,000 price tag on the project. Some of the senior managers found his ideas so innovative that they decided to make the firm a proving ground for a complete package, which they and the consultant would market to other firms. As the deal unfolded, the lawyers spent numerous hours studying the proposal, designing a plan to fit the strategy, and deciding how to finance the job. It looked so promising that the firm invested tens of thousands of dollars in the consultant's small company.

In the end, for unknown reasons, the consultant never delivered the goods. And, adding insult to injury, the lawyers soon learned that everything in the promised package was available off-the-shelf for half the price. "This happens all the time," says the former associate.

VENDORS

Size Up Tech Support

Because technical support varies so widely from product to product, it pays to shop around for the hardware and software that offer the best package. Your computer system is the heart of your company, and when it's down, you're down. Any vendor worth considering should offer **24-hour phone support, seven days a week**.

Some vendors charge for tech support, and some don't. In either case, be sure to find out whether a vendor you're considering has a toll-free number. You'll also want your vendor to be ready to provide you with overnight replacement parts for hardware. Many vendors now offer remote diagnostics, so they can diagnose a problem by dialing into your system right away, rather than sending someone to your site.

One hint from experts: Most vendors don't charge customers for advice until they have made a purchase or until their questions get technically complex. Translation: There's advice out there that's free for the taking.

VENDORS

Who Benefits from Telecom Reform?

The Telecommunications Act of 1996 signaled that both local and long-distance service were up for grabs. What's more, more phone companies would be fighting for the business. Here, experts answer questions people are asking about the new **telecommunications legislation and its potential to affect their companies**:

- *Will I get cheaper local service in the near future?*
  Don't hold your breath. Most fallout from the reform will take a while. "Expect five years of marketplace chaos and confusion," says Dave Goodtree, director of telecommunications strategies at Forrester Research, in Cambridge, Mass.
- *Is "one-stop shopping" for real?*
  That means a bundled package—local access, long-distance, and high-speed Internet access—most likely with a volume discount, handled by one vendor on one bill. "The good news is that you'll get more service choices; the bad news is that you'll have to pay for the advertising that tells you about it," says Bruce Egan, an affiliated research fellow at the Columbia University Center for Tele-Information.
- *What's the bottom line for small business?*
  "Businesses should say, 'Here are my problems as a small-business owner, give me a solution,'" advises Harris Miller, president of the Information Technology Association of America, in Washington, D.C. Play on providers' desire for long-term contracts by saying, "If you want my business for the long haul, here's what I want today."

VENDORS

Building a Web Site? Watch Your Wallet

Even the best informed small-business owners don't really know how to set themselves up on the Web. Prices vary wildly, and service providers aren't always what they seem. Here's how to avoid three of the most common costly mistakes:

- *Mistake 1: Hiring a high-priced designer.*
  The most expensive designer isn't necessarily the best. Shop around for an affordable yet skilled designer. Before you sign on with anyone, pay for a small mock-up of your site that shows its structure and illustrates the flow of visitor traffic. Make sure the designer understands your business and its message.
- *Mistake 2: Buying a special "Web server" computer.*
  Unless your site is video-intensive or links several databases and other functions to the site, you don't need more than a standard Pentium PC as your Web-site server. That's lucky for you. Specialized Web servers can cost up to $60,000. The real bottleneck for site traffic is likely to be the phone line, so invest in the capacity of your phone line and that of your Internet service provider.
- *Mistake 3: Getting lost in the hype.*
  Keep asking yourself, "How will this thing earn its keep? How will I measure effectiveness?" **Question every Web-building cost and expenditure**. Expect your site to change continually, and budget for that. But make sure you get your money's worth along the way.

• PURCHASING •

VENDORS

Early Birds Get the EDI Edge

"Our industry is old-fashioned; we tend to lag behind the times," says Michael Fidanza, general manager of Ideal Supply Co., a $20.5-million distributor of industrial pipes and valves in Jersey City, N.J. Still, when a top customer asked about the company's electronic data interchange (EDI) capability in 1990, Fidanza decided to take the plunge.

It's not at all unusual for large suppliers or customers to force a small company to install the hardware and software that makes EDI ordering possible. But Ideal made the move before that point and **installed EDI before its competitors**. By being one of the few EDI-ready buyers in its industry, Ideal was also well positioned to negotiate higher discounts.

"It's a good idea to get on board before your vendors and customers start demanding it," says Fidanza. The sooner an EDI partnership is in place, the more money and time both companies save. Ideal's vendors showed their appreciation. "One of our vendors offered us an extra $10,000 worth of product with every $50,000 purchase—just because we're EDI." Savings like that quickly covered Ideal's investment.

Ideal spent about $5,000 on EDI software and hardware that would cost $10,000 today. The service that processes Fidanza's purchase orders each month charges a low $80 per month. Now, up to 80% of Ideal's purchases and 15% of its sales are processed electronically.

VENDORS

E-mail or Else!

Tired of playing phone tag with vendors? Fed up with faxing your lawyer? Craig Aberle, president of MicroBiz, a software developer in Mahwah, N.J., was fed up. He issued an e-mail ultimatum to the nearly 1,000 people with whom he does business: **Get e-mail or get lost**.

About 95% of MicroBiz's $5 million in sales come through resellers across the country. Aberle sent each a letter saying, "We require all dealers to get e-mail, or we won't give you any leads." He threatened to stop doing business with those who refused to comply, figuring that "if only 50% of them agreed to go on e-mail, those are the 50% that are going to be in business next year." The same letter went to all of his less-dependent business partners who didn't have e-mail. His banker was already online.

At first there were grumblings, but within a few months, everyone was online. Since the mandate, MicroBiz's monthly phone bill has dropped from $35,000 to $16,000, and the resellers have started e-mailing each other.

VENDORS

Who Says You Can't Survive EDI?

Even before Jody Kozlow Gardner and Cherie Serota knew what it was, they dreaded electronic data interchange (EDI). Belly Basics, in New York City, is their maternity-clothing start-up, and the two women counted Federated Department Stores among their largest customers. When Federated mandated EDI, Belly Basics had six months to comply.

Ignoring EDI and dropping Federated wasn't an option for the $2.5-million company, which gets 25% of its orders from Federated. "Whatever it took, we had to do it," says Serota. EDI requires suppliers to receive computerized purchase orders from the retailer and return invoices in the retailer's chosen electronic form. Belly Basics' old modemless PC wasn't up to the task. Gardner and Serota started researching.

First, they realized that **being small gave them an advantage**. Their line was limited to about 20 items, so such tasks as computerizing inventory items wouldn't take much time. And since the company had already been putting UPC bar codes on the products, it had already met one EDI mandate.

The biggest challenge was order processing. The partners had two options: process the orders in-house with a $10,000 software/hardware package or hire a third-party service firm. Outsourcing won. "If 100% of our business were with Federated, we might bring EDI in-house," Gardner says. "But for two purchase orders a month, it's much better to go with a service." They found the service by asking Federated for recommendations.

In the end, they met the deadline, and Gardner admits, "The whole thing sounded a lot scarier than it is."

VENDORS

Who Needs EDI?

If you would like to **learn about electronic data interchange** (EDI) and the way your company might be affected, get a copy of *The Why EDI Guide for Small and Medium-Sized Enterprises* (EDI World Institute, 514-288-3555, $34.95). The book answers questions about paperless communications with vendors and customers, explaining how to exchange sales orders, product specs, invoices, and e-mail, electronically. The text is clear and free of technobabble and jargon.

The 108-page book includes case studies of 12 small companies—ranging in size from $3 million to $36 million—that implemented EDI. It reviews their reasons for adopting the system as well as the costs, structure, problems, and rewards. There is also a detailed questionnaire to help you determine whether your company should pursue EDI, and a companion worksheet highlights the expenses and benefits you can expect along the way.

VENDORS

Be Your Own Expert

Jim Gonyea found it tough going when he decided to set up his company's Web site. It seemed impossible to find reliable Internet expertise. "It's almost like the used-car business," he says. "You can easily go out and spend thousands of dollars and still not end up with a Web site."

The president of Gonyea & Associates, in New Port Richey, Fla., initially chose a consultant whose own storefronts were widespread on the Web. But the programmer assigned to Gonyea's project quit within four weeks, and four weeks later the second programmer quit as well. After spending several thousand dollars and, even worse, wasting four months, Gonyea gave up on the development firm.

Frustrated by the experience, Gonyea became **his own Web expert**. He hired two programmers to help him set up his own T1 phone line connection to the Internet. A hassle? It certainly was, Gonyea says, but it was worth it. "It'll take us more time and more money, but in the long run, we'll have more control," he says. "And that's the bottom line: You've got to get control of your data."

VENDORS

Ask, and Ye *Might* Receive

What can a small business expect from a large telecommunications company? Dave Goodtree, director of telecommunications strategies at Forrester Research, a business research firm in Cambridge, Mass., suggests a few **provisions a business should seek** as competition develops in the new telecommunications marketplace. You may not get everything, but you should ask.

- *Installation waiver.* This provision gives companies a way to offset the installation costs they incur every time they make changes to their telecommunications network. A business can request that a percentage of the contract's total dollar amount be set aside to take care of installation fees.
- *Revenue commitment.* To avoid getting locked into products and services that might soon become outdated or inappropriate, companies should commit to annual purchases worth a predetermined dollar amount rather than buying specific services.
- *Renewal bonus.* Carriers will consider giving bonuses to customers that renew contracts early or without opening them to outside bidders. The bonus can be as high as 10% of the contract's value and often comes in the form of a credit on a future bill.

VENDORS

Finding the Right Internet Provider

An Internet service provider (ISP) is an outfit with a direct link to the "backbone," or primary communication infrastructure, of the Internet. ISPs attach file servers and modems to their network links and sell Internet access to users, who connect through modems. If your Web site is hosted by an ISP, you don't have to worry about the technical issues and security concerns of managing the site. And not having those worries is a good part of why you pay for an ISP.

So, how do you **find a good ISP**? A good place to start looking is in your own backyard. Service providers usually advertise in local computer publications or in the business section of newspapers. And if you're able to browse the Web, you can find a comprehensive list of ISPs throughout the country at several sites, including www.thelist.com and www.clarinet.com/iap/iapcode.htm.

VENDORS

Six Routes to the Right ISP

Here are **six questions to ask when you're selecting an Internet service provider** (ISP):

1. *Does the ISP provide access to such basic services as e-mail, newsgroups, and file transfer protocol (FTP)?* If you're going online solely to get customer feedback, e-mail alone may be enough. But if you want to ship reports to satellite offices, your provider will have to offer FTP.
2. *How good is its technical support?* Your business must have someone who understands your particular needs. If technical service can't help you, don't bother.
3. *Does it offer the bandwidth, or speed of Internet access, that your business needs?* If you have large groups of employees who'll be using the connection simultaneously, you'll need high bandwidth. And your ISP should itself have a high-speed—a T1 or T3, for example—connection to the Internet.
4. *Is the price right?* Most services charge monthly fees that range from $10 to $40. Expect to pay more if you need high bandwidth and more than just basic service.
5. *How many POP—remote connection—sites does it have?* If you or your employees do a lot of business travel, a nearby POP gives Internet access for the price of a local phone call.
6. *Can it help you set up and maintain a Web site?* Some ISPs will only "host," or store the site. If yours doesn't offer design expertise, ask for a referral.

VENDORS

One-Stop Tech Shopping

Deciding where to buy hardware and software can pose overwhelming alternatives. Catalogs, specialty stores, and discount superstores abound. But despite the temptation to pinch pennies on mail order, it might be smarter for a small business to stay with one small local vendor, says Bob Olsen of Peregrine Outfitters, in Williston, Vt.

The outdoor accessory wholesaler started by **making all hardware and software purchases through one small vendor.** "That keeps things simple," says Olsen. "Let them decide whether I'm having a hardware or software problem." The vendor effectively acts as Olsen's information systems manager, supplying all his hardware and software.

Olsen has stuck with the same vendor for years. Now, with sales reaching $9 million, Peregrine is beginning to outgrow that vendor. Olsen knows that within the next couple of years, he'll have to make a major hardware and software upgrade, and at that time he'll need to move to a different vendor.

• ON THE ROAD •

"The thing...that I find fascinating is that we are not at a new place [with technology]. It's just becoming harder and harder to avoid the place we are."

NEWT GINGRICH
Speaker of the House of Representatives
Washington, D.C.

CELLULAR

Wirelessly Connected

In 1981, Steven Ettridge founded his temporary-employment agency, Temps & Co., in Washington, D.C. As the company grew in the mid-1980s, he needed a cellular phone to stay in contact with the office while he was on the road. Now Ettridge relies on his Porsche Carrera and its plug-in port for faxes and e-mail. The **car is his mobile office** when he shuttles among the company's 16 locations along the Washington-Baltimore-Philadelphia corridor.

Ettridge gets the most out of his mobile phone and pager, and he maintains "hot backups" of both: His car phone and mobile phone share one number, and he always keeps a spare pager at work.

Without his pager, Ettridge feels vulnerable. Cellular phones, he says, make "terrible incoming devices." Give out your number and not only will the thing ring constantly, but you're the one who's obliged to answer it. So Ettridge uses his phone almost exclusively for outgoing calls, and he responds to pager messages that he reads at his convenience.

The tiny pager is a real workhorse. Ettridge uses it to write himself messages, sort of like a string around the finger. He'll sometimes "call himself" and leave a message or a to-do list. That way his pager can remind him of important meetings or to make a critical phone call. And he sleeps with the pager on his bedside table, set for two wake-up buzzes. "It's my power alarm," he says.

CELLULAR

Battling E-mail Excess

The push-button ease of e-mail and the never-out-of-touch capabilities of a pager-cellular phone combo have many CEOs wondering how they managed without them. But like most other advances in productivity, they have their dark side. "The potential pitfall with e-mail is overload, and the same goes for pagers," says Gene Griessman, a time-management consultant in Atlanta, and author of *Time Tactics of Very Successful People* (McGraw-Hill, 800-262-7429, $14.95). "**Burnout is a major possibility if you're always accessible**. My recommendation: Put yourself in your appointment book. Schedule some time every day for quiet or recreation." Griessman offers four suggestions for keeping e-mail under control:

1. For intracompany e-mail, assign each message a priority ranking—A, B, C or 1, 2, 3—to flag its importance.
2. Preface each long message with a synopsis. The abstracts that precede academic articles are good models.
3. Also at the top of your messages, highlight upcoming action items.
4. Consider establishing a secondary e-mail account for urgent and high-priority messages or key clients.

CELLULAR

Shop for Cell Service

Don't fall for the free-cellular-phone pitch. To get the **most cost-effective cellular-phone service**, you need to ferret out the best package for your needs. Only then should you look for the free phone that goes with it. You can use a phone you already own. Coming to a deal with your own hardware should have no effect on the price of service.

Shop around for your local service, advises Atlanta-based Jeffrey Kagan, author of *Winning Communication Strategies* (Aegis Publishing, 401-849-4200, $14.95). Most users fall into two categories, says Kagan. Heavy users are on the phone more than 10 hours a month; light users are on less than an hour a month. Place yourself and start digging for deals.

Kagan suggests contacting resellers who buy service in bulk and offer it at discounted rates. MCI Cellular and AT&T Cellular, two of the largest resellers, offer nationwide service. But you shouldn't restrict yourself to considering only jumbo-size service providers in large cities. Most small providers located just outside large cities operate under the umbrella of large parent companies. They have the larger company's marketing, brand name, and expertise behind them, but many of them charge lower rates. Check local newspaper ads and the Yellow Pages, or call your local Chamber of Commerce.

For long-distance service, use your company's long-distance carrier and add the cellular service to your existing account. That way, your business discounts will automatically be applied to your cellular service. You can expect to save an average of 25%.

PAGERS

Get More from Your Pager

"My pager keeps me in touch with the world," says George Silva Jr., who runs G. J. Silva and Son Dairy, in Turlock, Calif. With 6,000 head of cattle and a 30-employee business to run, the owner of the $1-million dairy farm can't be tied to his office.

Silva, his two managers, and six other employees carry Motorola's Advisor Gold FLX pagers. For less than $120 a month, he leases all nine from a local office of PageNet, the nation's largest paging provider. Back in the office, company secretaries monitor the PCs and phones, and, when important calls come in—from a major customer like Safeway supermarkets, for example—they use the paging service to send e-mail messages to Silva and his managers. Silva relies on **three extra pager features**:

- *Scheduling*. To ensure that the farm's backup power supply is ready for emergencies, Silva has arranged for barn managers to receive this message every Monday at 11:30 a.m.: "Start generators before lunch. Shut them down after lunch."
- *E-mail*. Because his local PageNet office couldn't give him an e-mail address, Silva uses a third-party service.
- *News and weather*. "Weather is critical to my business," Silva says. "I don't have time to sit and watch it on TV." Throughout the day, his third-party e-mail service sends weather reports and national news updates. It also provides stock prices and sports scores.

PAGERS

Always on Call

Do you need a pager if you're already lugging around a cell phone and a laptop? Yes. For one thing, when you're on the road, it's often impossible to stop, plug in your computer, and download e-mail. For another, there are times when answering your cell phone is downright rude. "You never know just what you'll be doing when your phone rings," says Tim Scott, president and founder of MLC Mortgage, in Oklahoma City. "You might be on the golf course, admiring your client's swing. And interrupting that swing could cost you the deal."

Pagers alert you by vibrating quietly when they receive and store a new message. You **read or act on stored messages only when you're ready**.

Scott founded his six-employee mortgage lending firm in 1995. By keeping his overhead to a minimum, Scott can offer lower interest rates and fewer points than the big banks and financial institutions that are his competition. Still, he and his colleagues, who spend much of their time in the field, must follow up every lead to succeed. Because they know they have to return incoming calls promptly, he and his loan officers depend on pagers. When home buyers or real estate brokers call, the office manager uses a PC to page them, relaying messages like "New client. Call Sam Parker at 555-1212."

Do-It-Yourself Efficiency Expert

Karen Settle, president of Keystone Marketing Specialists, in Las Vegas, has mastered multitasking. Keystone, a $5-million company, provides employee computer training. Settle relies on about 300 independent contractors, and she uses conference calls to motivate and train them. During one such call with 30 contractors, Settle might hit her phone's mute button so that she can check her e-mail while still monitoring the conference.

"Because start-ups tend to run skinny on people, and there are only so many hours in the day, you have to do a lot more or you won't survive," Settle says. "I'm always looking for new **technology to increase productivity**."

Thanks to her pager, Settle is always available—emergency or not. Settle admits that she's often "overbeeped" by her staff. To combat that problem, she has demonstrated examples of unnecessary pages and urged employees to think carefully before calling her. Her lone guideline: If it's a client issue, use the pager.

Similarly, Settle has devised a strategy for dealing with the growing problem of too much e-mail. To avoid wasting the time it takes to scroll through junk e-mail and trivial messages, she has created an e-mail subaccount. She treats that account like an unlisted phone number, sharing its address only sparingly. Knowing that all e-mail to that account is coming from people she considers important, she is diligent about checking it frequently. An assistant monitors her general account, which Settle peruses herself when she has time.

PAGERS

Getting Paged, Discreetly

Keeping in touch is a two-way process. Not only do travelers have to maintain contact with their companies, their companies have to stay in touch with them. Pagers and cellular phones work, but they can be annoying and intrusive at times. The O.J. Simpson trial, for example, was enough of a circus without beeping pagers and ringing cellular phones.

So Judge Lance Ito, like most of his colleagues, banned their use in court. How did famed trial lawyer F. Lee Bailey stay in touch with the outside world? He depended on a Socket PageCard (Socket Communications, 800-552-3300). It's a pager on a PC card, and it works by itself, with a personal digital assistant, or with a laptop computer.

When it's plugged into a computer, a message window pops up whenever a page is received. Bailey, who had never used a pager before the Simpson trial, described it as "extremely useful." He explained that "judges usually permit computers in court, and it allows you to **receive messages without disrupting the proceedings**."

Managing Voice Mail

Fred DeLuca, who founded the Subway sandwich-shop chain, headquartered in Milford, Conn., can have any high-tech toy he wants. But his **most valued communications tool is humble voice mail**. Here's why:

"We started using voice mail more than 10 years ago. I like it because you don't have to have any special equipment. You see a phone; you make a call.

"I get about 60 messages a day from employees and franchisees, and I listen to all of them. For my sanity, I set a time limit of 75 seconds on my box, because people can be long-winded when they're excited. When I hear 'You have 30 messages,' I know right away that I'll spend 60 minutes on voice mail. I take two minutes per message, listening and returning or forwarding. And I'm addicted to using it in the car instead of turning on the radio. You can't do that with e-mail.

"Voice messages have more texture—expression and emotion—than e-mail or a memo does. Some people aren't readers; they're talkers, and you just can't capture them in writing the way you can in speech."

VOICE MAIL

Exploit Your Voice Mail

For the last 12 years, Richard Gordon has relied on voice-mail technology to help him oversee worldwide operations of his $60-million company. The president of R.j. Gordon & Co., a consulting firm with headquarters in Los Angeles, **checks his voice-mail box every 10 minutes**, forwarding and responding to messages from the more than 200 employees, customers, and vendors who are on his Meridian Mail system. To make voice mail work for you, he suggests the following:

- *Set up voice-mail boxes for important clients and vendors.* Gordon has boxes for about 30 clients and vendors. When they call the box, they hear, "You have a message, please enter your password to hear it."
- *Remind people how the system can help them.* Gordon's telecommunications manager regularly sends out bulletins to remind users of powerful features.
- *Save on long-distance calls.* When Gordon is overseas, he dials a toll-free number to access his voice-mail box. When he enters 0-9 before a U.S. phone number, Meridian Mail connects him at U.S. rates.
- *Reach key personnel in an emergency.* If a problem arises, an employee leaves a voice-mail message marked "remote notify urgent" in the emergency contact person's box. The phone system dials several preprogrammed numbers (home, cell phone, vacation site, overseas office, and so forth), leaving a message at each one.
- *Use broadcast lists sparingly.* Five or six is Gordon's limit, otherwise, "trying to remember who's on each one gets confusing."

ONLINE

Get Everyone Together—Electronically

In the old days, Katherine Emery held weekly technical meetings. Because the consultants at her 11-employee computer-support company work primarily in their clients' offices, Emery wanted to make sure they had opportunities to share ideas. So each Friday at Walker Systems Support, in Southington, Conn., consultants met for technical problem-solving sessions with their peers.

Now, Emery has a better way. Instead of returning to Walker Systems' offices for protracted meetings, her consultants **conduct their technical discussions online**. Using a shared, loosely structured database created with Lotus Notes software, consultants get help from their far-flung colleagues without leaving their clients' buildings. A Notes database can be managed on a company's internal Web site, or intranet, and remote workers can log in using a modem.

Walker Systems used to keep manila files bulging with customer information, storing everything from on-site visit reports to phone logs. Now, all that information is entered online, and it's easily available to any authorized user, no matter where he or she is working.

ONLINE

Options for Connections

When researching phone-line options for Internet access or other communications, it's easy to get lost in a sea of jargon. Here's a list of **phone-line options for digital service**:

Type of service	Speed	Line cost per month	Internet cost per month
Standard phone line	33.6 Kbps	$20-$60	$20-$40
ISDN	64 Kbps-128 Kbps	$70-$120	$200
Switched 56	56 Kbps	$70	$250
256-Kbps frame relay	56 Kbps	$300	$400
T1 frame relay	1.544 Mbps	$400	$1,200
Cable modem	Up to 500 Kbps	N.A.	$40

Prices are approximate and vary greatly depending on location and service details.

ONLINE

Office Hours on the Web

Brooks Mitchell has moved beyond the limits of e-mail and voice mail when he's not in the office. He used Netopia Virtual Office to set up his own mailbox on his company's Web site. Mitchell's SHL-Aspen Tree Software, in Laramie, Wyo., is a 35-employee recruiting-software company. He explains how the Web's **virtual office helps him stay in touch**:

"Anyone anywhere on earth who has a modem and Web-browser software now has real-time access to me through my computer. I made it all happen by following the software's step-by-step installation, which created a graphics-rich Web page.

"In effect, Virtual Office lets my computer act like a Web-site mini-server, and other Internet users can access the files that I've designated. I set security levels for people to access my virtual office. That protects the integrity of my files. At the site, visitors find up-to-date sales brochures, price lists, and other product information, and I receive messages from customers and progress reports from employees.

"Being able to communicate in absentia is a plus for businesspeople who, like me, often find themselves away from their desks. Whereas e-mail requires me to know an address for every person with whom I'd like to communicate, now I simply convey messages from my own Web site. No matter where I am, I regularly check and revise my site. Consequently, the messages I exchange with people are more informed than e-mail or voice mail could ever be."

ONLINE

Hotels Wired for Business

When David Peck checked into the Nob Hill Lambourne, in San Francisco, he expected to find a comfortable bed and maybe a few chocolates on his pillow. Peck, president of Reelin' in the Years Productions, a video archive company in San Diego, needed a place to stay while Reelin' was filming a testimonial for a band. What he got, along with the usual amenities, were free use of a state-of-the-art laptop computer and access to the Internet. These days, an increasing number of **hotels offer high-tech goodies** to business travelers.

The Nob Hill Lambourne hotel was the first San Francisco hotel to offer Internet connectivity in all its rooms. In 1989, the hotel outfitted its 20 rooms as business centers with desktop computers. After guests started to ask about laptop availability and Internet access, the hotel removed the antiquated PCs and invested $15,000 in new technology. Many other hotels have followed suit. So, when you make travel reservations, try to arrange for a room with a separate data phone line located near the desk.

Management by Modem

You can **manage home-based employees** with electronic bulletin-board systems (BBSs). Unibase, in Sandy, Utah, uses electronic bulletin boards to assign projects to its 2,000 employees and to monitor their progress. As a result, the $50-million company's four BBSs are essential to day-to-day internal operations. Here's how the setup works:

Unibase sells data-entry services to other companies, particularly corporations that collect information on handwritten forms but store their data on computers. Customers use Unibase-owned equipment to scan completed forms into a computer and to modem them—often cross-country—to Unibase. There, software programs organize the incoming documents into bundles of about 100 and distribute them over the company BBSs to Unibase employees, many of whom work at home at night.

Employees download files of the forms and key the handwritten data from the scanned forms into a typed format, ready for the customers' computer databases. After the employees finish typing the data, they again log into the BBS to return the files. Meanwhile, the BBS software enables managers to monitor the progress of assignments.

MOBILE COMPUTING

Cybersecretary

If you're your only employee, who minds the home office while you're away? Does your business receive telephone calls, faxes, and electronic files? You *could* organize a complicated series of call-forwarding and remote-communications systems to take over every time you leave the office. *Or* you could just program your trusty computer to handle everything while you're gone, giving clients the continuity of one point of contact for your company.

That's what Frank Boss does at Power Find, a start-up marketing-information service in El Cajon, Calif. "The **computer takes over my office while I'm gone**," he explains. "It receives calls, lets me know who called and when, and, while I'm away, I can remotely access my files." Boss uses BitWare (Cheyenne Software, 800-243-9462).

When callers dial his office, they can leave a message, send a fax, or upload and download data files. Boss has set up firewall protection to limit access to certain sections of his computer. His computer pages him whenever he has a message or fax, and Boss can phone in to retrieve the electronic missives whenever it suits him. He can arrange to have faxes forwarded to a nearby fax machine or to his laptop computer.

"I can retrieve anything from my office," he says. Boss is out of his office for 50 hours of his 80-hour workweek. So he's grown to depend on his technological sidekick.

MOBILE COMPUTING

Work Your Way Out of the Office

Many CEOs who work long, hard hours dream of cutting back. Although it may be a while before you can reduce your hours at work, even today you can **let technology help you prepare for a future when you can be outside the office**. Back in 1983, when John Sobeck started his company, First General Services of Northeastern Pennsylvania, in Wilkes-Barre, Pa., he worked 80 hours a week. But he had a vision. By age 55, he promised himself, he'd be his own man. Today, at 54, Sobeck spends only 20 hours a week at his $2-million fire-and-water-damage restoration business.

Sobeck's years of preparation paid off. Early on, he wrote policies-and-procedures manuals that help new workers learn their jobs quickly. He customized a computer program that holds all employees accountable for their work by keeping tabs on their activity. And every year he invests from $20,000 to $30,000 in technology that tracks the company's pulse. "The business could pretty much run by itself," Sobeck says.

His company continues to grow. He and some partners opened a field office in 1996. Even so, Sobeck continues to spend his summers fishing in North Carolina. And when he's in town? He hits the links by 2 p.m.

TELECOMMUTING

Hands-Free Travel Computing

Personal information managers are all the rage, but how do you use one of those tiny keyboards when you're driving? Jeffrey Epstein has found a better alternative: He relies instead on a miniature high-tech voice recorder. He's got all the other staples: cell phones, two-way pagers, and a palmtop. But his Voice Organizer 5500 (Voice Powered Technology, 800-255-2310) is at the top of his list.

"The recorder comes in handy for listening to directions as I'm driving to someone's house or office," he says. "I've had it for a while, and I'm still finding new ways to use it. It's **like having a penless notepad** wherever I go."

His Voice Organizer is palm-sized and shaped like a guitar pick. "I like to put it on my desk by my pager so that I look like a gadget guy," says Epstein, CEO of Boss Systems, a reseller of sales-force automation products in Chicago.

The tiny gizmo comes with 512 KB of memory, and Epstein uses it to record messages, schedule appointments, and store up to 100 phone numbers. To activate most functions, Epstein just pushes a button and speaks into the microphone. To retrieve a phone number, for example, he manually selects the letter of the alphabet where he stored a number. Then he speaks the name, and, instantly, it appears with the number on the liquid crystal display.

Hard Copy on the Go

Making sales presentations usually involves some sort of visual aid. Many techno-savvy salespeople rely on presentation software like Microsoft's PowerPoint. But what if you're on the road, and your prospect asks for printouts of the material? For highway hard copy, it's worth **investing in a travel printer**, says one CEO. Jim Noble of Noble Oil Services, in Sanford, N.C., bought a portable, three-pound Canon bubble-jet printer.

"My company recycles oil for its clients, which include the state of North Carolina and Jiffy Lube. I'm often on the road giving presentations to clients and potential clients about why one oil-disposal method is better than another. I have found that clients usually understand pictures far better than they understand words. With my travel printer, I can create professional color graphs and spreadsheets with speeds that rival my desktop printer (about four pages a minute). I can also print on acetate if I'm going to use an overhead projector during a presentation. In the past, if I needed something on the fly, someone back in my North Carolina office would have to send it to me."

TELECOMMUTING

Well-Timed Meetings

Jim Dieroff **takes his travel alarm clock** with him wherever he goes—even if it's just to a nearby meeting. Dieroff is president of Connaissance Corp., in Fort Collins, Colo., a $2-million consulting, continuing-education, and product-marketing business aimed at serving dental professionals. He originally bought the clock just for travel, but now he carries it everywhere he goes—to the office, to client locations, and on errands.

He finds that the small clock is especially useful at corporate meetings. "I just set the clock in front of me on the conference table," Dieroff says. When the rest of the meeting participants see the clock, they understand that Dieroff is serious about schedules and won't waste anybody's time. Also, Dieroff explains, a nonchalant glance at a watch during a meeting can seem rude and inappropriate. "This way, I never get in trouble for looking at my watch."

Computer Addiction

Slowly and subtly, it becomes a consuming habit. At first, you're online only occasionally, paying your bills, and maybe checking with your office once or twice. But then its lure grows more powerful. Soon you find that you're ignoring your family, spending all your spare moments in front of the screen. Laptops and cell phones make staying connected easier than ever, but it becomes hard to "Just Say No."

Fear not, computer junkie. Behavior therapists are standing by, and they're ready to help you shake off your cyberspace dependency.

Maressa Hecht Orzack, Ph.D., is founder and coordinator of Computer Addictive Services at McLean Hospital, in Belmont, Mass. "If the problem is an addiction to computer games, the cure could be as simple as deleting the offending programs from the hard drive," she says. "But if someone spends hours arranging his files, therapy might call for his **using a stopwatch to set a limit on the amount of time** he allows for the task."

The syndrome, Orzack cautions, is only going to spread. With more and more employees telecommuting, the line between work and play has blurred, opening the door to lengthening sessions on the Internet and longer stints of Solitaire during business hours. Schools are bringing younger kids online, making for ever-growing numbers of folks with hyperlinks on their mind. "People often gravitate toward the computer when they want to kill time or avoid an unpleasant task," says Orzack. "The problem is, nobody's teaching them when it's time to stop."

TELECOMMUTING

On the Spot—but Miles Away

When one of August Fromuth's top salespeople announced his intentions to **move to a distant part of the country while keeping the local territory** that he'd cultivated for two years, Fromuth, CEO of AGF Direct Gas Sales, in Manchester, N.H., gave Salesman X (who prefers to remain nameless) six months to show uninterrupted sales effectiveness. "Customers want to buy from a local rep," Fromuth warned X.

Five years later, Salesman X still lives afar, is still a top performer, and remains local in his customers' eyes. How does he maintain his salesman-next-door image? He relies on a Pentium laptop with internal fax/modem, a fax-receiving unit, several phone lines, voice mail, and a cellular phone. The fax-receiving unit pages X whenever a fax comes in. He downloads the fax by connecting the Fax Friday to his computer, which he has programmed so that it types a phone number from his territory on all his outgoing faxes.

The physical location of his phones is similarly camouflaged. For a $12 monthly service fee, his local phone company forwards all calls placed to his remote phone numbers to an 800 number, which then routes them to his office, cellular phone, or vacation spot.

X logs on to online services and phones news stations' update lines to keep on top of weather, sports, and news in his territory. "Weather is the most important," he says. "Selling natural gas, I need to know about a cold front up there when it's 80 degrees here."

TELECOMMUTING

Keep Telecommuters in the Fold

VeriFone, an electronic-payment processing company in Redwood City, Calif., is now a large, wholly-owned subsidiary of Hewlett-Packard, but since its earliest days, cofounder William Pape has followed the same management guidelines for **monitoring remote employees**. He recommends these strategies:

❧ *Visit your remote offices frequently*. Or, if your employees work out of their homes, have them come into the office for regular team or group meetings, say, once weekly or for one week out of every six to eight weeks.

❧ *Publish written guidelines that detail how you want your employees to set up their remote work spaces.* People at home need one separate, dedicated phone line for work. It should have its own answering machine or voice mail. And each home office should also have at least one other line for e-mail and faxes. Make sure that workers take breaks. It's easy for a home-based worker to spend too much time working without a recess.

❧ *Be sure your senior managers are themselves comfortable with remote communication.* Otherwise, they'll never comprehend the operating issues that confront remote workers.

❧ *Help remote workers form strong ties to people at the central office.* Any face-to-face meeting is an opportunity for cross-fertilization and to make sure that home-based employees feel a sense of belonging. Be more aggressive than centralized companies about scheduling time for socializing.

❧ *Find ways to compensate for the loss of daily face-to-face contact.* Encourage remote workers to e-mail or videoconference among themselves.

❧ *Set policies for when employees should hold face-to-face meetings.* Relying on e-mail can escalate minor irritations into major conflicts.

TELECOMMUTING

Finding the Best Telecommuters

Studies show that virtual companies—companies that use technology to link a dispersed staff—are showing productivity increases as high as 15%. How do companies **select and hire virtual workers**? You'll want people with the following traits:

- *Problem solving*. It's not easy for managers to know whether specific telecommuters are thrashing around unproductively on a problem. So you want people who are comfortable tackling problems independently.
- *Company loyalty*. Because they don't have daily face-to-face contact with you, it's more difficult for telecommuting employees to develop and sustain strong company ties. Testing for congruity with your company's values can be as simple as asking candidates to describe their ideal employer.
- *Strong work ethic*. Look for people who show a quick understanding of what needs to be done, who stay on task, and who deliver results, not excuses.
- *Self-confidence*. Working alone, it's easy to start second-guessing decisions and feedback. Look for people who feel secure about their job skills and personal lives.
- *Good sense of humor.* A good sense of humor can help employees-at-a-distance deal with the frustrations of a virtual workplace.
- *Technology-troubleshooting capabilities*. When they're home alone, remote workers have to feel confident about dealing with technical problems. Fortunately, that is a skill you can teach.

TELECOMMUTING

Laptops Accelerate Sales

At Gulf Industries, a sign-making company based in Torrance, Calif., president Kozell Boren purchased his first laptop computer several years ago to help him with sales. He installed Corel Draw (Corel Corp., 800-772-6735).

After a few months of tinkering with the laptop system, Boren handed it to salesman Perry Powell and told him to **use the laptop on sales calls**. The length of Powell's sales calls quickly shrank from three hours to one hour, and within a few months he became the company's number one salesperson.

Boren, who had seen revenues stay flat for five years, would have loved all 200 of his salespeople to follow Powell's lead, but he couldn't afford to buy a laptop for each of them. And, because the reps operate as independent contractors, Powell couldn't *force* them to lay out their own money. So he embarked on a major high-tech campaign, and little by little he converted half the reps. The hundred who held out maintained they had no need of laptops.

Within three years of launching the laptop program, company sales had increased by 29%. Those reps who had invested in the technology reported average sales increases of 25%. Those who shunned the computers can't help noticing that 45 of the 50 reps who won trips to Hawaii for having hit or exceeded their monthly quotas were laptop users.

L

M

N

O

P

R

S

T

U

V

W

X

Y

Z

Other business books from *Inc.* magazine

HOW TO *REALLY* CREATE A SUCCESSFUL BUSINESS PLAN
HOW TO *REALLY* CREATE A SUCCESSFUL MARKETING PLAN
HOW TO *REALLY* START YOUR OWN BUSINESS
By David E. Gumpert

MANAGING PEOPLE
HOW TO *REALLY* RECRUIT, MOTIVATE, AND LEAD YOUR TEAM
Edited by Ruth G. Newman
with Bradford W. Ketchum, Jr.

HOW TO *REALLY* DELIVER SUPERIOR CUSTOMER SERVICE
Edited by John Halbrooks

THE SERVICE BUSINESS PLANNING GUIDE
THE GUIDE TO RETAIL BUSINESS PLANNING
By Warren G. Purdy

ANATOMY OF A START-UP
WHY SOME NEW BUSINESSES SUCCEED AND OTHERS FAIL: 27 REAL-LIFE CASE STUDIES
Edited by Elizabeth K. Longsworth

MANAGING PEOPLE: 101 PROVEN IDEAS FOR MAKING YOU AND YOUR PEOPLE MORE PRODUCTIVE
FROM AMERICA'S SMARTEST SMALL COMPANIES
Edited by Sara P. Noble

301 GREAT MANAGEMENT IDEAS
FROM AMERICA'S MOST INNOVATIVE SMALL COMPANIES
Edited by Leslie Brokaw

301 DO-IT-YOURSELF MARKETING IDEAS
FROM AMERICA'S MOST INNOVATIVE SMALL COMPANIES
Edited by Sam Decker

301 GREAT CUSTOMER SERVICE IDEAS
FROM AMERICA'S MOST INNOVATIVE SMALL COMPANIES
Edited by Nancy Artz

www.inc.com/products

To receive a complete listing of *Inc.* business books and videos, please call 1-800-468-0800, ext. 5505.
Or write to *Inc.* Business Resources, P.O. Box 1365, Dept. 5505, Wilkes-Barre, PA 18703-1365.